Cities and the Global Politics of the Environment

"Starting with its title, *The Power of Cities in Global Climate Politics* never stops showing us how they are one of the key platforms for addressing climate change and for launching major transformations that go well beyond the city."
—Saskia Sassen, *Columbia University, USA*

"*The Power of Cities in Global Climate Politics* considers how and why innovation by city governments and their networks have such importance for climate change adaptation and mitigation. But also how much more needs to be done for them to get more substantial political power within their nation and on global climate politics and funds."
—David Satterthwaite, *International Institute for Environment and Development, UK*

More than half of humanity lives in cities, and by 2050 this might extend to three quarters of the world's population. Cities now have an undeniable impact on world affairs: they constitute the hinges of the global economy, global information flows, and worldwide mobility of goods and people. Yet they also represent a formidable challenge for the 21st Century. Cities are core drivers not only of this momentous urbanisation, but also have a key impact on the environment, human security and the economy. Building on the Palgrave Pivot initiative, this series aims at capturing these pivotal implications with a particular attention to the impact of cities on global environmental politics, and with a distinctive cross-disciplinary appeal that seeks to bridge urban studies, international relations, and global governance. In particular, the series explores three themes: (1) What is the impact of cities on the global politics of the environment? (2) To what extent can there be talk of an emerging 'global urban' as a set of shared characteristics that link up cities worldwide? (3) How do new modes of thinking through the global environmental influence of cities help us to open up traditional frames for urban and international research?

More information about this series at
http://www.springer.com/series/14897

"Cities wield increasing influence in climate governance, but key questions remain. What are the sources of their power? How effective can they be in mitigating or adapting to climate change? Can they be game-changing actors, shifting the norms that shape global politics? There is no better guide to these questions than Craig Johnson."

—Noah Toly, *Wheaton College, USA*

"Cities have made their way to the central stage of global climate governance and are arguably one of the most powerful forces for action. This book moves beyond existing accounts of the emergence of urban climate action to offer a powerful account of the possibilities and limits of cities acting on climate change and the consequences for how we must think about power and politics in international affairs."

—Harriet Bulkeley, *Durham University, UK*

Craig A. Johnson

The Power of Cities in Global Climate Politics

Saviours, Supplicants or Agents of Change?

palgrave
macmillan

Craig A. Johnson
University of Guelph
Guelph, ON, Canada

Cities and the Global Politics of the Environment
ISBN 978-1-137-59468-6 ISBN 978-1-137-59469-3 (eBook)
https://doi.org/10.1057/978-1-137-59469-3

Library of Congress Control Number: 2017952840

Printed on acid-free paper

This Palgrave Macmillan imprint is published by Springer Nature
The registered company is Macmillan Publishers Ltd.
The registered company address is: The Campus, 4 Crinan Street, London, N1 9XW, United Kingdom

To my mother, my children and the memory of my father.

"We do not inherit the Earth from our ancestors, we borrow it from our children."

—Indigenous proverb

Series Editors' Preface

The popularity of cities, as sites and as key actors in global environmental governance is perhaps one of the most defining themes of the past few years. From the stairs of Paris' City Hall, crowded with hundreds of mayors and celebrities on the eve of the Paris Agreement on climate change, to the collaborative response of city leaders the world over to nationalistic withdrawals against multilateral environmentalism, a global enthusiasm for the city is by all means a feature of our time. Yet many questions, as to the effectiveness, shape and legitimacy of cities in charting the future of world politics remain abundant and pressing.

Craig Johnson's *The Power of Cities in Global Climate Politics*, comes at a fruitful moment, and calls for closer scrutiny when it comes to the international-urban juncture. It offers us both a historical review of the role of cities in a set of particular geopolitical moments, as well as a clear assessment of their relationship with questions of climate adaptation and mitigation.

The book provides a solid discussion of the ways in which the emergence of transnational city-networks can inform how we understand power and politics in global climate politics. In the spirit of this book series, Johnson's book addresses explicitly the interplay between urban matters and global political themes, reminding us that much is still unanswered on the role of the city vis-à-vis international relations, and that the powers of mayors in shaping global governance might often be not quite what media attention and emerging rhetoric ascribe. Critically, the volume is a must read on the theme as it tackles nearly uncharted terrain, offering a thorough look into how the growth of urban action on climate change links to questions of power in global environmental regimes.

The productive road charted by Johnson in this volume is no doubt one worth following: as he flags in the conclusion of *The Power of Cities in Global Climate Politics*, there needs to be more attention to charting empirical, theoretical and normative pathways for further research, beyond rhetoric and conscious of history, as to the active presence of cities in world affairs. As to whether cities, mayors and urban actors might be the deciding factor in global environmental politics, the jury is—Johnson reminds us—still very much out. In a fine example of pragmatic and evidence-based research, with a healthy pinch of scepticism and some trust in the value of cities, *The Power of Cities in Global Climate Politics* opens debates and roads for further inquiry which we very much hope to continue in the volumes to come.

London, UK

Michele Acuto
Elizabeth Rapoport
Joana Setzer

Acknowledgements

In *The Death and Life of Great American Cities*, Jane Jacobs once remarked that "cities have the capability of providing something for everybody, only because, and only when, they are created by everybody." While this book probably won't provide something for everybody, it certainly bears the mark of many whose wisdom and inspiration have influenced my thinking about the power of cities in global climate politics. First and foremost, I'd like to thank Michele Acuto for encouraging and inspiring me into taking this project. I can think of few editors/colleagues whose humour, insights and patience made writing such a pleasure—when so often it is anything but! I'd also like to thank Sofie Bouteligier, Elizabeth Rapoport, and especially, Joana Setzer for their excellent comments and suggestions on previous drafts. I'm very grateful to Jack Redden and Rachael Ballard at Palgrave/MacMillan, Parthiban Gujilan Kannan at Springer Nature, and two anonymous reviewers for their comments, questions and very timely reminders about those ever-pending deadlines. My thinking about cities and city-networks would have been greatly diminished without the collaboration and friendship of Dave Gordon, whose insights about transnational climate governance and city-network orchestration had a profound impact on my thinking about city power and climate change. I'm also very grateful to Jen Allan, Melissa Ball, Arpana Chakravarty, Surma Das, Tom Deligiannis, Steffi Hamann, Iftekharul Haque, Kristy May, Matt McBurney, Sivhuoch Oh, Lauren Sneyd, and especially, Yvonne Su, for their truly inspirational work on the politics of urban and environmental sustainability in the Global South. Within my own department, I'd like to thank Jordi Díez, David MacDonald, Tim Mau, Andrea Paras, Bharat Punjabi, Adam Sneyd, and Ian Spears for their words of wisdom, encouragement and advice. I'm also very grateful to Harriet Bulkeley, Saskia Sassen, David Satterthwaite and Noah Toly for reading (and providing some kind and flattering comments on) an early version of the book, and to Mercy Borbor Cordova and Paty Romero-Lankao for some truly inspirational discussions about the future of urban sustainability at Habitat 3 in Quito. Finally, I'd like to thank Rosario Gómez for encouraging, inspiring and helping me to see the true meaning of performativity. Your words of support mean more than I can ever express, but one day I'll practice what I preach! This book is dedicated to my parents and to my children, but it extends to everyone whose fortunes are affected by the actions and decisions that we make—or fail to make—today.

Guelph, ON, Canada
October 2017

Craig A. Johnson

Contents

1

Introduction: The Power of Cities in Global Climate Politics

In order to fulfil my solemn duty to the United States and its citizens, the US will withdraw from the Paris climate accord…I was elected to represent the citizens of Pittsburgh, not Paris.

US President Donald Trump, 1 June 2017 (https://www.theguardian.com/environment/2017/jun/01/donald-trump-confirms-us-will-quit-paris-climate-deal. Last accessed 2 June 2017.)

As the Mayor of Pittsburgh, I can assure you that we will follow the guidelines of the Paris Agreement for our people, our economy & future.

Bill Peduto, Mayor of Pittsburgh, later that same day (https://twitter.com/billpeduto/status/870370288344674304. Last accessed 2 June 2017.)

Abstract This chapter introduces the volume by exploring the recent groundswell of enthusiasm surrounding the growing prominence of cities and city-networks in global climate politics. It first highlights the principal observations that have been made about the role of cities in global climate politics, highlighting the tensions and debates that are now animating the field. It then provides a roadmap for the rest of the volume, outlining its contribution to our understanding of city involvement in global climate politics.

C.A. Johnson, *The Power of Cities in Global Climate Politics*, Cities and the Global Politics of the Environment, https://doi.org/10.1057/978-1-137-59469-3_1

Keywords Cities · City-networks · Global climate governance
The politics of climate change

1.1 Introduction

The preceding statements by American President Donald Trump and Bill Peduto, the Mayor of Pittsburgh, capture a theme that now pervades the study of global climate politics. The first is that nation-states (and elected leaders of nation-states) have a "solemn" right to defend their citizens from international agreements and "entanglements" that undermine their national economic and political interests. The second is that cities (and their elected representatives) have a right and responsibility to act on climate change.

Twenty years ago, the idea that a mayor and an elected head of state would be locking horns over their commitment to climate change would have seemed bizarre, to say the least. But in 2017, city leaders around the world are now speaking and acting in the name of the planet—as well as their citizens. According to ICLEI, one of the world's largest transnational city-networks:

> In the United States already, 78 city and state government entities, representing almost 28 million US citizens, are monitoring their emissions reduction efforts through the carbon*n* Climate Registry. They are contributing to a global commitment to reduce emissions by more than one gigatonne of carbon-dioxide equivalent by 2030 — roughly the same amount pledged by the United States in its Nationally Determined Contribution to the Paris Agreement.[1]

In 2017, another city-network representing more than 80 of the world's largest cities, "the C40", responded to the US government's decision to withdraw from the Paris Climate Agreement by issuing a petition

[1]http://citiscope.org/commentary/2017/06/despite-trump-withdrawal-cities-and-states-will-ensure-climate-action-moves?utm_source=Citiscope&utm_campaign=80c9d5b717-Mailchimp_2017_06_02&utm_medium=email&utm_term=0_ce992dbfef-80c9d5b717-118068273. Last accessed 2 June 2017.

(that was signed by more than 50 city mayors, including the mayors of Accra, Amman, Paris and Toronto), calling upon the G20 heads of state to deliver on their Paris commitments to tackle climate change:

> As C40 mayors we will continue to lead on climate action in the most important cities of the world, standing for our people, the planet and global prosperity. Today, we seek to strengthen a pragmatic and positive alliance with you, in the service of our citizens. We look forward to working with you.[2]

Elsewhere, cities like Portland, Oslo, Medellín and Seoul are pursuing highly ambitious agendas for reducing emissions and vulnerability to climate change. According to the Non-State Actor Zone for Climate Action (NAZCA), another repository of information that is operated and maintained by the United Nations Framework Convention on Climate Change (hereafter UNFCCC), cities, sub-national governments, regions, investors, companies and civil society organizations accounted for a total of 11,615 climate change "commitments" in 2016.[3]

In the words of Seoul's mayor, said Park Won-soon, "local governments are actually leading national governments. They are the driving force" (in the global fight against climate change).[4]

But how do we make sense of these "forces?" And what do they tell us about the contemporary nature of international power?

For some (e.g. Tavares 2016; Barber 2013, 2017a), the growing prominence of cities in global climate politics suggests a fundamental transition from the old pattern of state-centric, multilateral governance that underlies the UNFCCC to a transnational, transformational arrangement that is rooted in the active involvement of sub- and non-state actors, including cities (e.g. Hale 2016). For others (e.g. Sassen 2015; Hickmann 2016; Gordon and Johnson 2017), the apparent transformation adds a layer of complexity and uncertainty to the study of global climate politics, suggesting the need for new theories

[2]https://www.change.org/p/g20-leaders-are-you-ready-to-save-our-planet. Last accessed 4 July 2017.

[3]http://climateaction.unfccc.int/. Last accessed 10 May 2016.

[4]http://time.com/4140172/paris-cities-states-climate-change/. Last accessed 6 May 2016.

and concepts that may be used to understand this process. For others, still (e.g. Davis 2016) the appearance of cities in global climate politics is but a temporary phenomenon that reflects the peculiarities of this particular moment in history.

Whether the "struggle" to craft an effective response to climate change becomes a zero-sum game of actions and reactions between cities and nation-states, the apparent rise of cities suggests a new set of norms and standards that are now being used to define what constitutes city leadership (Acuto 2013; Gordon and Acuto 2015; Acuto and Rayner 2016). Perhaps, the most powerful sentiment of this kind comes from the late Benjamin Barber, whose posthumous editorial in the Guardian newspaper captures the normative and political zeitgeist of contemporary city power:

> Because urban citizens are the planet's majority, their natural rights are endowed with democratic urgency. They carry the noble name of "citizen", associated with the word "city". But the aim is not to set urban against rural: it is to restore a more judicious balance between them. Today it is cities that look forward, speaking to global common goods, while fearful nations look back. (Barber 2017b)

Underlying Barber's comments is a powerful assertion that protecting the global atmospheric commons is an essential part of what it means to exercise political authority, and that cities have a right and a responsibility to intervene in this regard.

However, much remains to be known about the long-term implications of city and city-network engagement in global climate politics. Are cities and transnational city-networks, for instance, driving a coherent agenda that will have a lasting effect on reducing emissions and vulnerability to climate change? Are they legitimate actors in global climate governance? Are they able to provide a meaningful alternative to the multilateral system of nation-states? Above all, how does the growing involvement of cities and city-networks in global climate politics affect our understanding of international power?

This book seeks to address these questions.

In doing so, it makes the case that cities have emerged as international actors in their own right, but that their agency has been framed and constrained by the ways in which national governments, multilateral institutions, transnational networks and multinational corporations constitute their behaviour. In what follows, I argue that the power of cities to act and effect change in global climate politics can be usefully framed in relation to four constellations of international power. The first is framed primarily in relation to the formal rules, norms and expectations that are created by states in the context international regimes, in this case, the UNFCCC. The second stems from the constellation of norms, knowledge, ideas and resources that manifest themselves in transnational city-networks. The third lies in the ability of cities to accumulate and attract the labour, resources and capital that enable them to act and exert power at a global scale. The fourth and final form of power stems from the norms (of standardization, classification and evaluation) that render cities observable and comparable in global climate politics.

Understanding the power of cities in global climate politics, I contend, entails an ability to frame and define the ontological nature of cities, to theorize their power of influence, autonomy and agency in international affairs, and to orient these insights into a comparative historical and empirical analysis.

1.2 The Scholarly Terrain

There is now a palpable sense of optimism about the role that cities and transnational city-networks are playing in addressing some of the world's most pressing environmental problems (Barber 2013, 2017a; Tavares 2016). Nowhere was this optimism more apparent than during the run-up to the 21st Conference of the Parties in Paris, where cities were widely portrayed in social and print media as innovators or saviours whose actions were instrumental in providing critical leadership in the global fight against climate change (Weiss 2015; Worland 2015).

At the international level, particular attention has been paid to the role of transnational city-networks, such as the Cities Climate Leadership Group (C40), ICLEI and the Global Covenant of Mayors, whose membership activities have entailed advocacy, awareness raising and the dissemination of norms, knowledge and resources aimed at reducing emissions and vulnerability to climate change (Toly 2008; Bulkeley 2010; Gordon 2013; Bouteligier 2015; Bulkeley et al. 2015; Gordon and Acuto 2015; Johnson et al. 2015; Lee 2015; Hickmann 2016; Gordon and Johnson 2017). According to the carbonn Climate Registry (Deng-Beck and van Staden 2015), 608 cities, representing 553 million people, have registered commitments for reducing the equivalent of 1.0 GtCO2e, while NAZCA—the Non-State Actor Zone for Climate Action established by the UNFCCC Secretariat—reported 11,615 climate change commitments in 2016. Elsewhere, the C40 and Arup (Arup-C40 2015a) report that "228 global cities, representing 436 million people, have set greenhouse gas reduction goals and targets amounting to a cumulative reduction of 13 GtCO2e by 2050".

At the heart of this transformation is a recognition that global climate governance has shifted away from purely multilateral governance arrangements (where authority derives primarily from the power of nation-states) to a hybrid of transnational (Abbott 2013) and polycentric (Ostrom 2010) governance arrangements, in which a much larger range of actors is now shaping (or at least trying to shape) the "global climate governance landscape" (Betsill et al. 2015). For many cities, Kyoto was often invoked as a source of inspiration (or despair, as the case may be) that urban leaders could use in framing their own climate change initiatives (Bulkeley 2010; Bulkeley and Betsill 2013; Burch et al. 2015; Johnson et al. 2015; Setzer et al. 2015). Indeed, there is now a large body of evidence that many cities used the language of the Kyoto Protocol and of climate change more generally to justify new forms of policy and investment at the urban scale (Bulkeley 2010; Bulkeley and Castan Broto 2012; Bulkeley and Betsill 2013; Bulkeley et al. 2015; Toly 2008; Kern and Bulkeley 2009; Acuto 2013; Gordon 2013; Gordon and Acuto 2015; Gore 2015; Johnson et al. 2015; Lee 2015; Sassen 2015; Setzer et al. 2015; Gordon and Johnson 2017).

Concerns about the inability of national governments and international institutions to achieve meaningful cuts have therefore reinvigorated discussions about the role of cities and transnational city-networks in filling the gap (Tavares 2016; Barber 2017a, b).

Yet, amidst the euphoria, there is also a sense that the power that has been ascribed to—and frequently assumed by—cities has been overstated; that the power of cities to make a difference in global climate politics is not what it appears. Although many cities are now "speaking the language" of climate change, the ability of cities, city leaders, planners and politicians to implement policies that reduce emissions and vulnerability is often highly dependent upon the administrative channels that govern a wide range of sectors, including transportation, water and sanitation, health, housing and emergency services (Acuto 2013; Revi et al. 2014; Aylett 2015; Gordon and Acuto 2015; Johnson et al. 2015; Lee 2015; Setzer et al. 2015; Hickmann 2016; Gordon and Johnson 2017). Indeed, the factors affecting the viability and effectiveness of urban climate policy initiatives (e.g. trade policies, globalization, food and fuel subsidies) are often well beyond the power of any single municipality or local authority (Sassen 2015).

Moreover, the impact of cities and city-networks on urban and international politics remains poorly understood, reflecting the myriad ways in which cities and other urban interests may conceivably interact with global climate policy networks and processes (Bouteligier 2015; Gordon 2013; Gordon and Acuto 2015; Gore 2015; Lee 2015; Johnson et al. 2015; Romero-Lankao et al. 2015; Setzer et al. 2015). By and large, the vast majority of writing about city engagement in transnational city-networks has focused on the efforts of large, industrialized cities to reduce (and advance an international agenda for reducing) GHG emissions (Acuto 2013; Bulkeley and Betsill 2013; Gordon 2013; Gordon and Acuto 2015; Gordon and Johnson 2017). To date, far less attention has been paid to the politics of urban climate policy formation in the Global South (Carmin et al. 2012; Gordon 2013; Gore 2015; Johnson et al. 2015; Lee 2015; Leichenko 2011).

Recent empirical work on the factors affecting transformative climate governance has shown that cities can provide important sites of experimentation, in which new plans, codes and policies have been used to reduce emissions and vulnerability at the urban scale (Hoffmann 2011; Bulkeley and Castan Broto 2012; Bulkeley et al. 2015; Anderton and Setzer 2017). Bulkeley et al. (2015: 19) define urban climate policy experiments as "purposive and strategic" interventions that are (1) open-ended about the possible range of impacts and outcomes; (2) aimed at reducing GHG emissions or impacts; and (3) "delivered by or in the name of an existing or imagined urban community". By way of example, they include cases of zero carbon housing, solar thermal heating and other efforts whose impact on GHG emissions and impacts is strategic, yet uncertain.

An important point (of enthusiasm) that emerges from the literature on urban climate policy experiments is that cities can serve as "policy incubators" for testing new approaches, documenting possible outcomes, sharing best practices and envisioning alternative policy futures (Boyko et al. 2012; Bulkeley et al. 2015; Anderton and Setzer 2017). Framed in this way, urban policy experiments provide an important means of challenging existing norms, practices, interests and investments that underlie ecologically destructive path dependencies (Geels 2004; Geels and Raven 2006; Geels and Schot 2007). However, questions can be raised about the extent to which the forces of experimentation and innovation are dependent upon the knowledge, networks and capital that tend to concentrate in relatively affluent urban centres (Florida 2017).

Sustainability transition theories suggest that relatively isolated changes and events can lead to wider systemic changes when they disrupt the dominant sociotechnical and socio-ecological systems that underlie the creation of goods and services, the consumption of resources and the production of waste and pollution (Geels 2004; Geels and Raven 2006; Geels and Schot 2007; Bulkeley et al. 2015). Within this literature, important distinctions are made between "*technological niches*", spaces of innovation where new and potentially "unstable" forms of experimentation occur; "*sociotechnical landscapes*", the exogenous environment in which macro-economic, cultural and macro-political developments shape human actions and decisions; and "*sociotechnical regimes*", the broad constellation of norms, regulations, interests and investments that stabilize and support existing path dependencies (Geels and Schot 2007: 400).

According to Geels and Raven (2006), sustainability transitions occur when niche experiments are able to reconfigure broad sociotechnical landscapes, thereby destabilizing the rigidity and resistance of existing sociotechnical regimes (Geels 2004; Geels and Raven 2006; Geels and Schot 2007). Clearly, transitions theories have strong bearing on the study of urban climate experiments, but they have also been criticized for overstating the transformative potential of technological niches and for understating the agency that critical stakeholders (i.e. corporations and foreign capital) have in effecting sociotechnical change (Berkhout et al. 2004; Smith et al. 2005). Within the urban context, for instance, questions have been raised about the extent to which urban policy experiments are dependent upon the availability of wealth, resources and external investment (Bulkeley et al. 2015).

In short, there is now a growing recognition that cities and city-networks are playing a critical role in governing the politics of climate change. However, much remains to be known about the ways in which and extent to which cities and city-networks are making a discernible difference in mitigating and adapting to climate change. Are cities, for instance, driving a coherent agenda that can facilitate the transition to a low-carbon economy? Or are they simply responding to the norms and practices put in place by national governments and international institutions? Are cities and city-networks providing leadership and representation in multilateral and transnational climate governance? Or do they represent particular cities, groups and interests (e.g. land owners, corporations and political elites) that effectively exclude other important stakeholders, such as slum dwellers, informal labour and the urban poor? Finally, and most centrally, how does the rise of cities in global climate politics inform our understanding of cities, nation-states and international environmental politics?

1.3 Aims, Contributions and Rationale

The Power of Cities in Global Climate Politics is about the ways in which the growing prominence of cities and transnational city-networks in global climate politics informs our understanding of power and politics in the international system. Drawing upon the extant literature on transnational climate governance, climate mitigation and urban adaptation, it aims to stimulate new thinking about the ways in which scholars are conceptualizing the rise of cities, including the ways in which these insights inform our understanding of power and politics in global environmental governance regimes.

At first glance, the field appears inundated with books and edited volumes about cities, global environmental governance and climate change. Upon closer inspection, a case can be made that the attention that has been paid to the three fields of global climate governance, cities and climate change, and cities and global governance has largely occurred in isolation. Few books about global climate governance, for instance, pay explicit attention to the urban dimensions of mitigation and adaptation (e.g. Biermann and Pattberg 2012; Stevenson 2013; Stevenson and Dryzek 2014). Moreover, many of the books that deal most directly with cities and global climate governance are about Northern experiences in relatively wealthy industrialized countries (e.g. Bulkeley and Betsill 2003; Meyer 2013; Stone Jr. 2012), although Bulkeley et al. (2015) and van der Heijden (2017) are important recent exceptions.

Although there is no shortage of books and edited volumes about the role of cities in global climate governance (e.g. Johnson et al. 2015; Lee 2015; Tavares 2016; Barber 2013, 2017a), many of these are either empirical studies that generate new data about the factors and processes affecting urban and global climate governance (e.g. Tavares 2016) or normative ones that argue in favour of acknowledging or extending the role of cities in global climate affairs (Barber 2013, 2017a). To date, relatively little has been written about the ways in which city engagement in global climate politics affects our understanding of power in global environmental regimes.

In filling these gaps, the book pursues four interrelated aims: first, it brings together the latest theoretical and empirical work on the ways in which cities are now engaging in global climate politics at different scales of analysis; second, it aims to stimulate new thinking about the ways in which city engagement in global climate politics informs our understanding of power and politics in global climate governance; third, it compares the politics of mitigation and adaptation; and finally, it seeks to identify critical ways of thinking about the ways in which cities and city-networks may contribute to low-carbon development pathways that reduce emissions and vulnerability to climate change.

1.4 Framing the Analysis: The Power of Cities in Global Climate Politics

One of the challenges of theorizing the power of cities in global climate politics is that they lack the sovereignty that is typically afforded to nation-states (Lee 2015). Thinking about the role of cities in global climate politics therefore requires a theory of power that can account for the ways in which states and non-state actors affect the long-term patterns of continuity and change in international politics (Gordon and Acuto 2015; Acuto and Rayner 2016; Hickmann 2016; Gordon and Johnson 2017).

Theoretical approaches to the study of cities in global climate politics are relatively new, reflecting the speed and scale at which cities and city-networks have expanded over the last two decades (Gordon and Johnson 2017; Herrschel and Newman 2017; van der Heijden 2017). According to Bulkeley (2010), early research on the development of urban climate governance dates back to the mid-1990s, and there, the focus was on the internal dimensions of climate policy processes and decisions in primarily Northern industrialized cities. More recently, growing attention has been paid to the "external" orientation of cities, focusing primarily on the actions and decisions of global cities (Lee 2015), city-networks (Gordon and Acuto 2015; Hickmann 2016; Gordon and Johnson 2017), paradiplomacy (Tavares 2016) and international relations (Acuto 2013; Acuto and Rayner 2016; Herrschel and Newman 2017).

A common theme that pervades the study of cities and global climate politics is the notion that city engagement challenges or disrupts the prevailing Westphalian order of nation-states (Barber 2013, 2017a; Tavares 2016; Herrschel and Newman 2017). Within the field of international climate governance, particular attention has been paid to the role of cities and other sub- and non-state actors in "catalyzing" new agendas for reducing emissions and vulnerability to climate change (Chan et al. 2016; Hale 2016; Anderton and Setzer 2017; Bäckstrand et al. 2017; Bäckstrand and Kuyper 2017; Gordon and Johnson 2017; van der Ven et al. 2017; van der Heijden 2017). By and large, the growing presence of cities and city-networks is seen positively, but questions have been raised about the extent to which cities are able to chart a course that is autonomous and independent of that being pursued by multilateral institutions, multinational corporations and nation-states (Hickmann 2016; Gordon and Acuto 2015; Gordon and Johnson 2017; Herrschel and Newman 2017).

One popular metric that has been used to compare city efforts on climate change (Chaps. 3, 4) is the idea of counting and weighting policy actions and commitments that cities report to third party disclosure platforms such as NAZCA, carbon*n* and the CDP (formerly the Carbon Disclosure Project). To their credit, reporting platforms such as these provide a systematic means of compiling and comparing city actions and commitments on climate change. At the same time, they have been criticized for their relative lack of attention to political interests, processes (Davidson and Gleeson 2015) and policy outcomes (Widerberg and Stripple 2016; van der Heijden 2017).

A second way of understanding city power in global climate politics is to theorize the relative degree to which cities are able to advance a policy agenda that challenges or contradicts other agendas, including those of international institutions, multinational corporations and nation-states. In his recent book on authority in global climate governance, Thomas Hickmann (2016) conceptualizes city-network power in terms of conflictual, complementary and dependent relations, suggesting that city power can be discerned by documenting the extent to which cities and city-networks are resisting, complementing or relying on external institutions and forces. Similarly, the C40's Climate Action in Mega Cities Report, CAM 3.0, differentiates cities on the basis of whether and to what extent they are able to legislate, regulate and implement policy in the absence of private sector and other actors (Arup-C40 2015). Finally, *Deadline 2020*, also by the Arup-C40 (2016) classifies power in terms of (1) whether cities own and operate the assets and infrastructure that are necessary for implementing climate policy;

(2) whether they set and enforce their own policies and regulations; (3) whether they are able to control their budgets; and (4) whether they are able to set a vision for the city's climate future.

A third perspective suggests that cities can attract and project power by situating themselves within transnational networks that consolidate actors and resources for the purposes of adaptation and mitigation. As noted earlier, transnational city-networks like ICLEI and the C40 have come to play a critical role in pooling resources, sharing ideas and providing critical points of access to finance, technology and expertise used in the development or low-carbon and climate-resilient pathways (Toly 2008; Funfgeld 2010; Romero-Lankao et al. 2015; Gordon and Acuto 2015; Lee 2015; Setzer et al. 2015; Gordon and Johnson 2017). At the same time, city-networks depend on the financial and intellectual contributions of national governments, international donors and multinational corporations, suggesting that their ability to act autonomously in global climate politics may be constrained (Gordon and Acuto 2015; Hickmann 2016; Gordon and Johnson 2017).

If cities and city-networks are dependent upon external actors, how do we make sense of their power? Building upon the work of Ken Abbott et al. (2015) and others working on the politics of transnational orchestration (e.g. Hale and Roger 2014; Chan et al. 2015; Bäckstrand and Kuyper 2017; Widerberg 2017), Gordon and Johnson (2017) make the case that city-network power can be usefully understood as a kind of orchestration that entails city-networks serving as intermediaries in governing a target population (Abbott et al. 2015). In particular, they identify three different forms of orchestration: (1) *complementary orchestration*, in which city action in global climate politics is largely framed and constrained by the power of sovereign nation-states operating within the multilateral regime; (2) *concurrent orchestration*, whereby cities and city-networks are able to undertake coordinating actions and activities that produce collective effects that are not necessarily undertaken "with an eye to the interstate regime" (Gordon and Johnson 2017: 11); and (3) *emergent orchestration*, which implies that power and order are "by-products" of the technical standards, common methodologies and standardized metrics that cities choose to voluntarily adopt when engaging with transnational networks, multinational corporations and the UNFCCC.

Theories of network orchestration offer a potentially interesting lens for documenting and interpreting the kinds of soft power that are now being used to nudge cites and city leaders into acting on climate change (Hale and Roger 2014; Bäckstrand and Kuyper 2017; Gordon and Johnson 2017). However, questions can also be raised about the extent to which orchestration processes are leading to the consolidation as opposed to the fragmentation of city power (Gordon and Johnson 2017). First, conforming to network norms, standards and metrics provide a powerful means by which orchestrators (e.g. national governments, bond rating agencies and international donors) and intermediaries (e.g. ICLEI and the C40) can observe, compare and control the actions and decisions of cities, suggesting a loss of agency and autonomy. Second, and related, being observed, compared and evaluated facilitates access to external resources that can in theory be withheld or denied.

Questions about the relative autonomy of cities and city-networks raise a final set of questions about the ways in which climate policy norms and commitments are interpreted and acted upon at the urban scale (Bulkeley 2010; Chu et al. 2015; Castan Broto 2017). The empirical literature on urban climate policy formation suggests that cities are more likely to emulate and adopt international climate policy commitments when local politicians, city officials, experts and local communities are able to make a clear and credible connection between acting on climate change and generating tangible co-benefits, in the form of ambient air quality, affordable housing and sustainable livelihood (Bulkeley 2010; Aylett 2015; Bouteligier 2015; Burch et al. 2015; Gordon and Acuto 2015; Lee 2015).

Particularly important in this regard is the ability of policy actors and entrepreneurs (Toly 2008) to frame and re-frame policy discourses surrounding the local politics of global climate change (Bulkeley 2010; Aylett 2015; Bouteligier 2015; Burch et al. 2015; Anderton and Setzer 2017; van der Heijden 2017). According to Dryzek, policy discourses are "metaphors and other rhetorical devices that are deployed to convince listeners or readers by putting a situation in a particular light" (Dryzek 2013: 19). Politically and analytically, their importance lies in their ability to convey positive and negative perceptions about the legitimacy of particular groups, issues and policy agendas (Schön and Rein 1994;

Ingram et al. 2007; Dekker 2017), highlighting the idea that certain policies may flourish in an environment that lends a sense of urgency and legitimacy to a particular policy ideal (cf. Finnemore and Sikkink 1998). However, the ability of cities and city leaders to frame climate policy discourses as viable and essential is itself dependent upon the constellation of ideas, institutions and interests that affect the perceived viability and desirability of acting on climate change (Pierson 2004; cf. Davidson and Gleeson 2015; Aykut 2016; van der Heijden 2017).

According to Lukes (2005), power implies an ability to influence actions and agendas through decisions, non-decisions and ideological frames. The "first face of power", he argues, involves an explicit choice "among alternative modes of action" (Lukes 2005: 22), suggesting "a focus on behaviour in the making of decisions on issues over which there is an observable conflict of (subjective) interests" (Lukes 2005: 19). The second implies "a means by which demands for change … can be suffocated before they are even voiced; or kept covert; or killed before they gain access to the relevant decision making arena; or failing all these things, maimed or destroyed in the decision implementing stage of the policy process" (Lukes 2005: 22–23). Finally, a third entails the construction and possible manipulation of ideas, preferences and interests that are used in framing and legitimating particular policy actions and ideals (Lukes 2005).

As we shall see, the power of cities in global climate politics has assumed all three of these guises, providing alternative models of climate governance that are at times highly dependent upon the power of city-networks, multinational corporations and nation-states, and in different contexts liberating cities from the path dependencies of carbon-intensive maladaptive development. By suggesting that cities are "saviours, supplicants and agents of change", I make the case that cities are under certain circumstances able to wield and project considerable power by leading and influencing international climate policy agendas while in others, they are clearly dependents, whose need and demand for international assistance also constitutes a form of power. The third and final face of power is the idea that cities are also important constellations of actors, interests and institutions, whose underlying dynamics create the conditions for resilient, low-carbon development pathways that extend far beyond the urban scale.

1.5 Outline of the Book

So, how does the book proceed?

The following chapter sets the stage for the rest of the book by situating the rise of cities in relation to the study of international relations, international political economy and global environmental change. In so doing, it makes the case that the growing prominence of cities in global climate politics can be usefully understood in relation to four critical historical junctures:

- The end of the Cold War, which foreshadowed the conceptualization of non-state power, complex interdependence and multilevel governance;
- The globalization of information and communication technologies, which facilitated the rise of transnational networks, global cities and post-Fordist production regimes;
- the re-scaling of global production chains, reflecting the rise of the NICs, the BRICs and other emerging economies; and
- An emerging consensus that human activities have become the primary cause and determinant of global environmental change, introducing a new epoch of the Anthropocene.

After situating the rise of cities and city-networks in relation to these three historical/theoretical transitions, I then make the case that city power can be usefully understood by differentiating the terms on which cities engage with the international system, whose essential characteristics are conceptualized as follows:

1. **Westphalian power**—the notion that city power is framed primarily in relation to the formal rules, norms and expectations that are created by nation-states within the context of the UNFCCC;
2. **Network power**—the notion that city power stems from the constellation of norms, knowledge, resources and relationships that manifest themselves in transnational city-networks;

3. **Corporate power**—the idea that city power lies in the ability of cities to accumulate and attract the labour, resources and capital that enables them to act and exert power at a global scale;
4. **Performative power**—the notion that cities are able to project power by articulating, institutionalizing and normalizing a role for themselves within local and global policy discourses.

Although the ideas and assumptions being used to delineate these conceptual categories are by no means mutually exclusive, they do provide an important heuristic that may be used to pinpoint the different dimensions and mechanisms through which city power is being projected in global climate politics.

Drawing upon this framework, Chap. 3 then explores the ways in which cities have exerted and projected power in the context of global climate mitigation. Central to the analysis is an assertion that the agency of cities in global climate politics has been framed (and in many ways constrained) by the Westphalian power of the UNFCCC. By looking at a number of recent climate mitigation initiatives, the chapter explores the extent to which city-networks like ICLEI, the C40 and the Global Covenant of Mayors have been able to develop a viable alternative to the multilateral framework. Underlying the work of many city-networks is a corporate model of climate governance that entails disseminating and adopting particular standards, metrics and methodologies, suggesting new forms of performative power.

Chapter 4 next moves to the politics of urban climate adaptation, exploring the extent to which cities and city-networks have been able to articulate and negotiate new forms of power by demonstrating the need for the development of industries, assets, infrastructure and human settlements that will facilitate adaptation to climate change. Building upon the previous chapter, I make the case that the politics of urban climate adaptation have occurred in reaction and response to a Westphalian system that effectively reifies the power of nation-states. However, animating, disseminating and in some cases contradicting the politics of multilateral climate governance is a parallel process of network governance and MNCs, whose metrics, indicators and benchmarking exercises provide important forms of performative power that are being used to organize and orchestrate the ways in which cities, city-networks and corporations are defining resilience, vulnerability and risk in the Paris climate regime.

Finally, Chap. 5 concludes the analysis by highlighting the critical ways in which city engagement in global climate politics affects our understanding of international power. The implications for conceptualizing and theorizing the power of cities are considered, as are the methodologies that may be used in exploring and explaining the changing power of cities in global environmental governance regimes.

References

Abbott, K.W. 2013. Strengthening the transnational regime complex for climate change. *Transnational Environmental Law* 3 (1): 57–88.

Abbott, Kenneth W., Philipp Genschel, Duncan Snidal, and Bernhard Zangl. 2015. *International organizations as orchestrators*. Cambridge: Cambridge University Press.

Acuto, Michele. 2013. *Global cities, governance and diplomacy: The urban link*. New York and London: Routledge.

Acuto, M., and S. Rayner. 2016. City networks: Breaking gridlocks or forging (new) lock-ins? *International Affairs* 92 (5): 1147–1166.

Anderton, K., and J. Setzer. 2017. Subnational climate entrepreneurship: Innovative climate action in California and São Paulo. *Regional Environmental Change*. doi:10.1007/s10113-017-1160-2.

Arup-C40. 2015. *Climate action in megacities 3.0*. http://www.cam3.c40.org/images/C40ClimateActionInMegacities3.pdf. Last accessed 14 June 2017.

Arup-C40. 2016. *Deadline 2020: How cities will get the job done*. http://www.c40.org/researches/deadline-2020. Last accessed 14 June 2017.

Aykut, S.C. 2016. Taking a wider view on climate governance: Moving beyond the 'iceberg', the 'elephant', and the 'forest'. *Wiley Interdisciplinary Reviews: Climate Change* 7 (3): 318–328.

Aylett, Alex. 2015. Relational agency and the local governance of climate change: International trends and an American exemplar. In *The urban climate challenge: Rethinking the role of cities in the global climate regime*, ed. Craig Johnson, Noah Toly, and Heike Schroeder, 156–180. New York and London: Routledge Press.

Bäckstrand, K., and J.W. Kuyper. 2017. The democratic legitimacy of orchestration: the UNFCCC, non-state actors, and transnational climate governance. *Environmental Politics* 26 (4): 764–788.

Bäckstrand, Karin, Jonathan W. Kuyper, Björn-Ola Linnér, and Eva Lövbrand. 2017. Non-state actors in global climate governance: from Copenhagen to Paris and beyond. *Environmental Politics* 26 (4): 561–579.

Barber, Benjamin R. 2013. *If mayors ruled the world: Dysfunctional nations, rising cities*. New Haven, CT: Yale University Press.

Barber, B. 2017a. *Cool cities: Urban sovereignty and the fix for global warming*. New Haven: Yale University Press.

Barber, B. 2017b. How to fix climate change: Put cities, not countries, in charge. *The UK Guardian,* May 7. https://www.theguardian.com/commentisfree/2017/may/07/fix-climate-change-put-cities-not-countries-in-charge-oslo-seoul.

Berkhout, F., A. Smith, and A. Stirling. 2004. Socio-technological regimes and transition contexts. In *System innovation and the transition to sustainability: Theory, evidence and policy*, ed. B. Elzen, F.W. Geels, and K. Green, 48–75. Cheltenham: Edward Elgar.

Betsill, M., N.K. Dubash, M. Paterson, et al. 2015. Building productive links between the UNFCCC and the broader global climate governance landscape. *Global Environmental Politics* 15 (2): 1–10.

Biermann, Frank, and Philipp Pattberg (eds.). 2012. *Global environmental governance reconsidered*. Cambridge: MIT Press.

Bouteligier, Sofie. 2015. Multinational companies and urban climate governance: Market making or successful policy innovation? In *The urban climate challenge: Rethinking the role of cities in the global climate regime*, ed. Craig Johnson, Noah Toly, and Heike Schroeder, 82–100. New York and London: Routledge Press.

Boyko, C., et al. 2012. Benchmarking sustainability in cities: The role of indicators and future scenarios. *Global Environmental Change* 22: 245–254.

Bulkeley, H. 2010. Cities and the governing of climate change. *Annual Review of Environment and Resources* 35: 229–253.

Bulkeley, H., and M. Betsill. 2003. *Cities and climate change: Urban sustainability and global environmental governance*. London and New York: Routledge.

Bulkeley, H., and M. Betsill. 2013. Revisiting the urban politics of climate change. *Environmental Politics* 22 (1): 136–154.

Bulkeley, H., and V. Castan Broto. 2012. A survey of urban climate change experiments in 100 cities. *Global Environmental Change* 23 (1): 92–102.

Bulkeley, Harriet, Vanesa Castan Broto, and Gareth A.S. Edwards. 2015. *An urban politics of climate change*. London and New York: Routledge.

Burch, Sarah, Alison Shaw, Freya Kristensen, John Robinson, and Ann Dale. 2015. Urban climate governance through a sustainability lens: Exploring the integration of adaptation and mitigation in four British Columbian cities. In *The urban climate challenge: Rethinking the role of cities in the global climate regime*, ed. Craig Johnson, Noah Toly, and Heike Schroeder, 119–138. New York and London: Routledge Press.

Carmin, J., I. Anguelovski, and D. Roberts. 2012. Urban climate adaptation in the global south: Planning in an emerging policy domain. *Journal of Planning Education and Research* 32 (1): 18–32.

Castan Broto, V. 2017. Urban Governance and the Politics of Climate Change. *World Development*. http://dx.doi.org/10.1016/j.worlddev.2016.12.031.

Chan, S., H. Asselt, T. Hale, K.W. Abbott, M. Beisheim, M. Hoffmann, and P. Pauw. 2015. Reinvigorating international climate policy: A comprehensive framework for effective nonstate action. *Global Policy* 6 (4): 466–473.

Chan, S., C. Brandi, and S. Bauer. 2016. Aligning transnational climate action with international climate governance: The road from Paris. *Review of European Community and International Environmental Law* 25 (2): 238–247.

Chu, E., I. Anguelovski, and J. Carmin. 2015. Inclusive approaches to urban climate adaptation planning and implementation in the global south. *Climate Policy* 16 (3): 1–21.

Davidson, K., and B. Gleeson. 2015. Interrogating urban climate leadership: Toward a political ecology of the C40 network. *Global Environmental Politics* 15 (4): 21–38.

Davis, D. 2016. Can mayors actually rule the world? *Citiscope*. http://citiscope.org/habitatIII/commentary/2016/12/can-mayors-actually-rule-world?utm_source=Citiscope&utm_campaign=a10e62b5f0-Mailchimp_H3_2016_12_05&utm_medium=email&utm_term=0_ce992dbfef-a10e62b5f0-118068273. Last accessed 7 December 2016.

Dekker, R. 2017. Frame ambiguity in policy controversies: critical frame analysis of migrant integration policies in Antwerp and Rotterdam. *Critical Policy Studies* 11: 127–145.

Deng-Beck, C., and M. van Staden. 2015. *Carbonn Climate Registry Five Year Overview Report 2015*. http://www.carbonn.org. Last accessed 14 June 2017.

Dryzek, John. 2013. *The politics of the earth: Environmental discourses*. Oxford: Oxford University Press.

Finnemore, M., and K. Sikkink. 1998. International norm dynamics and political change. *International Organization* 52 (4): 887–917.

Florida, R. 2017. *The new urban crisis: How our cities are increasing inequality, deepening segregation, and failing the middle class—And what we can do about it*. New York: Basic Books.

Funfgeld, Hartmut. 2010. Institutional challenges to climate risk management in cities. *Environmental Sustainability* 2: 156–160.

Geels, F. 2004. From sectoral systems of innovation to socio-technical systems: Insights about dynamics and change from sociology and institutional theory. *Research Policy* 33: 897–920.

Geels, F. and R. Raven. 2006. Non-linearity and Expectations in Niche-Development Trajectories: Ups and Downs in Dutch Biogas Development (1973–2003). *Technology Analysis & Strategic Management* 18 (3–4): 375–392. doi:10.1080/09537320600777143.

Geels, F. and J. Schot. 2007. Typology of sociotechnical transition pathways. *Research Policy* 36: 399–417.

Gordon, David J. 2013. Between local innovation and global impact: Cities, networks, and the governance of climate change. *Canadian Foreign Policy Journal* 19 (3): 288–307. doi:10.1080/11926422.2013.844186.

Gordon, David, and Michele Acuto. 2015. If cities are the solution, What are the problems? The promise and perils of urban climate leadership. In *The urban climate challenge: Rethinking the role of cities in the global climate regime,* ed. Craig Johnson, Noah Toly, and Heike Schroeder, 63–81. New York and London: Routledge Press.

Gordon, D., and C. Johnson. 2017. The orchestration of global urban climate governance: Conducting power in the post-Paris climate regime. *Environmental Politics* 26: 694–714.

Gore, Christopher. 2015. Climate change adaptation and African cities: Understanding the impact of government and governance on future action. In *The urban climate challenge: Rethinking the role of cities in the global climate regime*, ed. Craig Johnson, Noah Toly, and Heike Schroeder, 205–226. New York and London: Routledge Press.

Hale, T. 2016. "All hands on deck": The Paris agreement and nonstate climate action. *Global Environmental Politics* 16 (3): 12–22.

Hale, T., and C. Roger. 2014. Orchestration and transnational climate governance. *The Review of International Organizations* 9 (1): 59–82.

Herrschel, T, and P. Newman. 2017. *Cities as international actors: Urban and regional: Governance beyond the nation state*. London: Palgrave/Macmillan.

Hickmann, T. 2016. *Rethinking authority in global climate governance: How transnational climate initiatives relate to the international climate regime.* London and New York: Routledge Press.

Hoffmann, Matthew J. 2011. *Climate governance at the crossroads experimenting with a global response after Kyoto.* Oxford: Oxford University Press.

Ingram, H., A. Schneider, and P. deLeon. 2007. Social construction and policy design. In *Theories of the policy process*, ed. P. Sabatier, 93–128. Boulder: Westview Press.

Johnson, Craig, Noah Toly, and Heike Schroeder. 2015. Urban resilience, low carbon governance and the global climate regime. In *The urban climate challenge: Rethinking the role of cities in the global climate regime*, 3–23. New York and London: Routledge Press.

Kern, K., and H. Bulkeley. 2009. Cities, Europeanization and multi-level governance: Governing climate change through transnational municipal networks. *JCMS: Journal of Common Market Studies* 47 (2): 309–332.

Lee, Taedong. 2015. *Global cities and climate change: The translocal relations of environmental governance*. New York: Routledge.

Leichenko, Robin. 2011. Climate change and urban resilience. *Current Opinion in Environmental Sustainability* 3: 164–168.

Lukes, Steven. 2005. *Power: A radical view*, 2nd ed. London: Palgrave Macmillan.

Meyer, William B. 2013. *The environmental advantages of cities: Countering commonsense antiurbanism*. Cambridge: MIT Press.

Ostrom, E. 2010. Polycentric systems for coping with collective action and global environmental change. *Global Environmental Change* 20: 550–557.

Pierson, P. 2004. *Politics in time: History, institutions and social analysis*. Princeton: Princeton University Press.

Revi, A., D. Satterthwaite, F. Aragón-Durand, J. Corfee-Morlot, R.B.R. Kiunsi, M. Pelling, D.C. Roberts, and W. Solecki. 2014. Urban areas. In *Climate change: Impacts, adaptation, and vulnerability*. Part A: Global and-Sectoral Aspects. Contribution of Working Group II to the Fifth Assessment Report of the Intergovernmental Panel on Climate Change, ed. C.B. Field, V.R. Barros, D.J. Dokken, K.J. Mach, M.D. Mastrandrea, T.E. Bilir, M. Chatterjee, K.L. Ebi, Y.O. Estrada, R.C. Genova, B. Girma, E.S. Kissel, A.N. Levy, S. MacCracken, P.R. Mastrandrea, and L.L. White, 535–612. Cambridge: Cambridge University Press, United Kingdom and New York.

Romero-Lankao, Patricia, et al. 2015. Multilevel governance and institutional capacity for climate change responses in Latin American cities. In *The urban climate challenge: Rethinking the role of cities in the global climate regime*, ed. Craig Johnson, Noah Toly, and Heike Schroeder, 181–204. New York and London: Routledge Press.

Sassen, S. 2015. Bringing cities into the global climate framework. In *The urban climate challenge: Rethinking the role of cities in the global climate regime*, ed. Craig Johnson, Noah Toly, and Heike Schroeder, 24–36. New York and London: Routledge Press.

Schön, D.A., and M. Rein. 1994. *Frame Reflection: Toward the Resolution of Intractable Controversies*. New York: Basic Books.

Setzer, Joana, Laura Valente de Macedo, and Fernando Rei. 2015. Combining local and transnational action in the adoption and implementation of climate policies in the city of Sao Paulo. In *The urban climate challenge: Rethinking the role of cities in the global climate regime*, ed. Craig Johnson, Noah Toly, and Heike Schroeder, 101–118. New York and London: Routledge Press.

Smith, A., A. Stirling, and F. Berkhout. 2005. The governance of sustainable socio-technical transitions. *Research Policy* 34: 1491–1510.

Stevenson, Hayley. 2013. *Institutionalizing unsustainability: The paradox of global climate governance*. California: University of California Press.

Stevenson, Hayley, and John Dryzek. 2014. *Democratizing global climate governance*. Cambridge: CUP.

Stone Jr., Brian. 2012. *The city and the coming climate: Climate change in the paces we live*. Cambridge: Cambridge University Press.

Tavares, R. 2016. *Paradiplomacy: cities and states as global players*. Oxford: Oxford University Press.

Toly, Noah J. 2008. Transnational municipal networks in climate politics: From global governance to global politics. *Globalizations* 5 (3): 341–356.

van der Heijden, J. 2017. *Innovations in urban climate governance: Voluntary programs for low-carbon buildings and cities*. Cambridge: Cambridge University Press.

van der Ven, H., S. Bernstein, and M. Hoffmann. 2017. Valuing the contributions of nonstate and subnational actors to climate governance. *Global Environmental Politics* 17 (1): 1–20.

Weiss, K. 2015. Cities bask in spotlight at Paris climate talks. *Nature*. http://www.nature.com/news/cities-bask-in-spotlight-at-paris-climate-talks-1.19006. Last accessed 10 December 2015.

Widerberg, O. 2017. The 'Black Box' problem of orchestration: how to evaluate the performance of the Lima-Paris Action Agenda. *Environmental Politics* 26 (4): 715–737. doi:10.1080/09644016.2017.1319660.

Widerberg, O., and J. Stripple. 2016. The expanding field of cooperative initiatives for decarbonization: A review of five databases. *Wiley Interdisciplinary Reviews: Climate Change* 7: 486–500.

Worland, J. 2015. *How cities and states took the spotlight in Paris climate talks time magazine*. http://time.com/4140172/paris-cities-states-climate-change/. Last accessed 5 July 2017.

2

Understanding the Power of Cities in Global Climate Politics: A Framework for Analysis

Abstract This chapter sets the stage for the rest of the book by situating the rise of cities in relation to the study of international relations, international political economy and global environmental governance. In so doing, it makes the case that the growing prominence of cities in global climate politics can be usefully understood in relation to four critical historical junctures: (1) the end of the Cold War, which foreshadowed the conceptualization of non-state power, complex interdependence and multilevel governance; (2) the globalization of information and communication technologies, which facilitated the rise of transnational networks, global cities and post-Fordist production regimes; (3) the re-scaling of global production chains, reflecting the rise of the NICs, the BRICs and other emerging economies; and (4) an emerging consensus that human activities have become the primary cause and determinant of global environmental change, introducing a new epoch of the Anthropocene. By situating the rise of cities and city-networks in relation to these four historical/theoretical transitions, this chapter outlines a conceptual framework for understanding the interaction of Westphalian, network, corporate and performative power.

C.A. Johnson, *The Power of Cities in Global Climate Politics*, Cities and the Global Politics of the Environment, https://doi.org/10.1057/978-1-137-59469-3_2

Keywords Theories of power · The power of cities · Westphalian power Network power · Corporate power · Performative power

2.1 Introduction

Prior to the 1990s, the vast majority of academic writing about cities and global climate politics had been limited to the study of individual cities taking what appeared to be relatively discrete actions on "climate-sensitive" sectors, such as transportation, waste disposal, housing and energy (Bulkeley 2010; Acuto 2013; Gordon and Acuto 2015; Gordon and Johnson 2017). By the turn of the century, cities and transnational city-networks had emerged as leading players in the global fight against climate change (Acuto 2013; Chan et al. 2015; Gordon and Acuto 2015; Hale 2016; Hickmann 2016; Gordon and Johnson 2017). In the words of Harriet Bulkeley:

> … urban networks have taken a more overtly political stance toward the issue, seeking to position cities as critical sites for addressing the issue of climate change or even opposing national governments (as in the United States and Australia), and in so doing have advanced claims for the strategic importance of urban governance. (Bulkeley 2010: 233)

Bulkeley (2010) identifies two historical "waves" of city engagement that help to explain the academic and political repositioning of cities within the field of global climate politics. The first, she argues, occurred in the early 1990s, signifying the convergence of the pioneering efforts on the part of primarily European and North American cities to address climate change and the mobilization of transnational networks (such as ICLEI, the Climate Alliance and Cities for Climate Protection) engaged in supporting and disseminating local climate governance experiments (cf. Hoffmann 2011; Bulkeley and Castan Broto 2012; Bulkeley et al. 2015; Gordon 2016; Hale 2016). The second entailed what she calls "a new generation of municipal networks and a more geographically diverse range of cities" (Bulkeley 2010: 232). According to Bulkeley (2010), it was this second wave of city mobilization that now marks an important shift from what was previously a relatively isolated and localized phenomenon to a wider systemic presence in global climate politics.

How do we explain this transformation? And what does it tell us about the power of cities in global climate politics?

This chapter sets the stage for the rest of the book by situating the rise of cities and city-networks—both historically and comparatively—in the study of global climate politics, thereby outlining a framework for analysis. It starts from the premise that cities operate in a wider political economy that shapes and constrains the ability of urban governance institutions to act and interact at the global level. In contrast to other policy fields, climate change governance entails the formulation, dissemination and interpretation of scientific knowledge (on climatic processes, changes and impacts) that highlights the role of epistemic communities and boundary organizations in framing and translating science into workable policy and practice. Finally, city engagement in global climate politics entails the coordination and cooperation of myriad state and non-state actors across multiple scales of analysis.

In the previous chapter, I made the case that the power of cities in global climate politics can be usefully understood as different forms of power:

1. **Westphalian power**— the notion that city power is framed primarily in relation to the formal rules, norms and expectations that are created by states within the context of the UNFCCC;
2. **Network power**—the notion that city power stems from the constellation of norms, knowledge, ideas, resources and relationships that manifest themselves in transnational city-networks;
3. **Corporate power**—the idea that city power lies in the ability of cities to accumulate and attract the labour, resources and capital that enables them to act and exert power at a global scale;
4. **Performative power**—suggests that cities are able to project power by articulating, institutionalizing and normalizing a role for themselves within the global environmental discourses that constitute the climate regime.

Each of these themes we now explore in more detail, thereby laying the groundwork for a conceptual framework that can be used in understanding the power of cities in global climate politics. The following section first considers the notion that city power derives primarily from the place they are able to establish for themselves within the multilateral climate regime, the UNFCCC. The second section then explores the idea that the power of cities stems from the constellation of norms, knowledge, ideas, resources and relationships that manifest themselves in the form of transnational networks. Section 2.4 theorizes the power of cities in relation to the processes of economic and political restructuring that have taken place in the context of liberalization, globalization and neoliberal reform. Finally, the last section outlines a performative theory of power that may be used to understand the ways in which cities are engaging with transnational networks, multinational corporations and the multilateral regime.

2.2 Westphalian Power: Cities and States in the Global Climate Regime

As noted in Chap. 1, one of the challenges in theorizing the power of cities in global environmental politics is that they lack the formal standing that is often accorded to sovereign nation-states within international regimes, such as the UNFCCC (Acuto 2013; Gordon and Acuto 2015; Curtis 2014; Lee 2015; Keohane and Victor 2011, 2016; Hickmann 2016; Herrschel and Neuman 2017). As Taedong Lee (2015: 32–33) has argued, cities are on the one hand bound by the rules of sovereign nation-states while, on the other, nested within policy networks (such as ICLEI, the C40 or the UNFCCC) that often transcend national boundaries and jurisdictions (Toly 2008; Gordon 2013, 2015; Gordon and Acuto 2015).

As we have seen in Chap. 1, many cities are now playing—or at the very least, claiming to play—a central role in the global fight against climate change. Underlying such claims is a wide range of actions, decisions and commitments that are designed to meet or exceed the goals and targets that have been agreed upon under the auspices of the UNFCCC. Prior to the Paris Climate Agreement, many cities used the goals and targets of the Kyoto Protocol to frame their own policy commitments, often in spite of their own national governments (Aylett 2015; Burch et al. 2015; Johnson et al. 2015). Following Paris, many others are now being recognized and registered under the auspices of the multilateral system (e.g. the Non-State Actor Zone for Climate Action—NAZCA).

Clearly, being recognized by or in relation to the UNFCCC conveys a certain degree of status that can conceivably be used in other domestic and international contexts. Whether cities are able to exert power in international climate politics depends on the ways in which national governments and international organizations frame and define the rules and decision-making procedures governing international regimes (e.g. the UNFCCC). According to Krasner (1982), regimes are the "...principles, norms, rules and decision-making procedures around which actors' expectations converge in a given area" (Krasner 1982: 186). Framed in this way, principles imply "beliefs of fact, causation and rectitude"; norms are "standards of behavior"; rules imply "specific prescriptions or proscriptions for action"; and decision-making procedures are the "prevailing practices for making and implementing collective choice" (Krasner 1982: 186).

According to Keohane and Victor (2011), states build international regimes as a means of realizing the benefits of cooperation with other states and with non-state actors, such as multinational corporations and NGOs. However, regimes vary in terms of the power, coherence and compliance they are able to achieve in relation to member states and relevant non-state actors. "At one extreme", they argue,

> ... are fully integrated institutions that impose regulation through comprehensive, hierarchical rules. At the other extreme are highly fragmented collections of institutions with no identifiable core and weak or nonexistent linkages between regime elements. In between is a wide range that includes nested (semi-hierarchical) regimes with identifiable cores and non-hierarchical but loosely coupled systems of institutions. (Keohane and Victor 2011: 8)

Whether regimes are able to achieve strong integration and compliance, they argue, is contingent upon (1) the interests of individual nation-states; (2) the material gains states can achieve through linkages; and (3) their ability to manage the uncertain actions, intentions and outcomes of other states, non-state actors and events (Keohane and Victor 2011). Focusing specifically on climate change mitigation, Keohane and Victor (2011: 7) make the case that "there is no integrated regime governing efforts to limit the extent of climate change. Instead, there is a *regime complex*: a loosely coupled set of specific regimes" (emphasis added).

Underlying these interpretations is an assumption that international regimes exist primarily as a result of states acting to create rules, order and predictability among individual nation-states, whose sovereignty varies in relation to the power and influence they are able to exert over their own populations, on international institutions and among other states (Keohane and Victor 2011). The implication here is that cities derive their power from the institutional place they are able to occupy within the multilateral climate regime (Gordon and Acuto 2015; Acuto and Rayner 2016; Gordon and Johnson, forthcoming). As Acuto and Rayner (2016: 1151) remind us, "City diplomacy is still deeply embedded in the Westphalian system".

Drawing upon Keohane and Victor (2011), we can theorize that city engagement in climate mitigation and adaptation policy processes will occur in two very different *regime complexes*, involving two different and distinct sets of interests, linkages and uncertainties about actions, intentions and outcomes. For mitigation, the primary concern is about assigning responsibility for reducing emissions and calculating the relevant costs and implications for those who are required to comply (Chap. 3). The interests, linkages and uncertainties are rooted in a "global" discourse about collective goods, consumption, burden sharing and the development of alternatives to fossil fuels. For adaptation (Chap. 4), the principal concern is about the terms on which eligible actors (principally countries) can legitimately claim and obtain access to a wide range of entitlements that may reduce vulnerability to climate change. Here, the interests, linkages and uncertainties are rooted in a discourse about vulnerability, impacts and international assistance in a (relatively) local context (Chap. 4).

Differentiating between mitigation and adaptation complexes provides an important means of understanding the role of cities in the global climate regime. Conceptually, it helps to distinguish between the formal power that local authorities are able to exercise in relation to zoning, congestion charging, etc. (these are primary questions of multilevel governance) and the wider scales that affect and are affected by resource needs and decisions at the urban scale. It also draws our attention to the translocal networks (Lee 2015) and phenomena that connect cities with national governments, international organizations' transnational networks and other cities (Toly 2008; Curtis 2014; Lee 2015; Sassen 2015). Finally, it makes us aware of the "place-making" (or "place-defending") politics of resistance that appear to be so pronounced in a context of hyper-connectedness (Harvey 1990; Castells 1996).

All of this being said, there is also a growing realization that a great deal of the action that is taking place in the name of climate change mitigation and adaptation is occurring outside of the official multilateral regime (Toly 2008; Bulkeley et al. 2015; Hale and Roger 2014; Betsill et al. 2015; Chan et al. 2015; Abbott 2017; Hale 2016; Shi et al. 2016; Gordon and Johnson 2017; van der Ven et al. 2017; Widerberg 2017), highlighting the role of city-networks in global climate politics.

2.3 Network Power: Cities as Non-State Actors

A second form of power thus focuses less on the institutions, interests and incentives governing the formal multilateral regime, and more on the ways in which norms and institutionalized practices order priorities, frame expectations and define what constitutes appropriate and acceptable behaviour within the international system. At the heart of this formulation is the notion that transnational networks confer new forms of power that enables cities to transcend the politics of multilateral governance (Lee 2015; Acuto and Rayner 2016; Hickmann 2016; Gordon 2016; Gordon and Johnson 2017). In Acuto and Rayner's words (2016: 1157), "The ways in which cities connect with and relate to each other, and trade (or indeed steal) ideas, are central to charting the physical shape of an increasingly urbanized age".

Theories of network governance suggest that norms and institutionalized relations can provide an important means of ordering priorities, framing expectations and defining what constitutes appropriate behaviour (Finnemore and Sikkink 1998; Bernstein and Cashore 2012;

McCann 2011; Paterson et al. 2014; Betsill et al. 2015; Lee 2015; Gordon and Johnson 2017). As Bernstein (2001: 5) has argued,

> Norms constitute identities and meanings by defining who may act, in what context they may act, and what their actions mean in that particular context. They regulate by pre/proscribing how actors should behave in defined contexts ... Finally, norms serve a deontic function when they express values that create rights and responsibilities and thereby empower actors by providing reasons or justifications for particular actions.

Finnemore and Sikkink (1998: 891) further distinguish between "*regulative norms*, which order and constrain behavior and *constitutive norms*, which create new actors, interests, or categories of action".

Through conferences, workshops and other forms of policy engagement, city engagement in transnational policy networks facilitates the diffusion of norms, knowledge and resources that alter the willingness and ability of cities and core urban constituencies to adopt, implement and monitor climate change commitments, metrics and actions for developing, measuring and tracking progress on emission reductions policies (Bouteligier 2015; Gordon 2013; Gordon and Acuto 2015; Gore 2015; Setzer et al. 2015; Hickmann 2016; Anderton and Setzer 2017; van der Heijden 2017). Cities also frequently use other cities or "early adopter cities" as "benchmarks" for their own policy innovations, suggesting that networks of city leaders, officials, experts and community representatives can foster networks for wider forms of adaptive learning and transformative change (Bulkeley and Castan Broto 2012; Carmin et al. 2012; Burch et al. 2015; Chu et al. 2015; Bulkeley et al. 2015; Lee 2015; Setzer et al. 2015; Hickmann 2016; Anderton and Setzer 2017; van der Heijden 2017). Finally, transnational city-networks provide an important means of pooling resources, sharing information and gaining critical points of access to finance, technology and expertise that might be used in developing low-carbon pathways (Toly 2008; Chu et al. 2015; Carmin et al. 2012; Gore 2015; Hughes and Romero-Lankao 2014; Funfgeld 2015; Romero-Lankao et al. 2015; Gordon and Acuto 2015; Lee 2015; Setzer et al. 2015; Hickmann 2016).

The implication then is that transnational city-networks provide important norms, knowledge and resources that cities may use in adopting international commitments, metrics and actions on climate change. However, the theoretical and empirical conditions under which local policy experiments may lead to wider transformations are by no means clear (Bulkeley et al. 2015). First, the vast majority of writing about urban climate policy experiments has focused on larger, wealthier cities, whose access to public finances, private investment and international climate policy networks makes them more visible in the extant literature on cities and climate change (Acuto 2013; Gordon 2013; Gordon and Acuto 2015). Second, questions can be raised about the challenge of conceptualizing cities and urban policy experiments in isolation from the wider systems of resource production and consumption in which they are embedded (Sasson 2015).

Underlying the idea that transnational networks may facilitate the diffusion of policy norms is an assumption that policies that are developed in one political jurisdiction can be recognized, emulated and adopted in other, comparable policy settings (Finnemore and Sikkink 1998; Paterson et al. 2014; Betsill et al. 2015). However, shifting the analysis from the urban to wider scales of analysis presents a number of ontological challenges. For one, the distinction between urban and non-urban is by no means crystal clear, reflecting the informality of land and resource governance that often occurs in and around large cities (Satterthwaite 2010; Saunders 2010; Leaf 2011). Second, the social, economic and cultural relations that connect cities with non-urban areas and communities suggest an ecological footprint that obscures the urban/rural dichotomy (Seto et al. 2012; Brenner 2013).

In short, cities may derive their power by engaging *directly* in the UNFCCC or they may derive it *indirectly* by engaging in transnational networks whose actions and authority derive from the multilateral regime. But what makes cities powerful and influential in the first place? From the introductory chapter, we can infer that city engagement in global climate politics varies enormously; many—but not all—of the cities that have been most active in global climate politics are large, "global" cities, whose economic, political and cultural clout enables them to occupy pivotal niches within the global climate policy landscape (Betsill et al. 2015). The following section next situates the study of city power in relation to the political economy of urban change.

2.4 Corporate Power: The Materiality of Urban Form

A third form of power stems from the material flows of resources, finances, labour and capital that affect the economic and wider political fortunes of cities, metropolitan regions and nation-states (Castells 1996; Friedmann 1986; Sassen 2001; Roy 2009, 2010, 2011; Webster 2011; Leaf 2011; Brenner 2013; Brenner and Schmid 2013; Florida 2017). In contrast to Westphalian and network theories of international power, corporate power rests primarily in the ability of cities to occupy strategic niches within the world economy, particularly in a historical context of globalization and neoliberal reform (Castells 1996; Friedmann 1986; Sassen 2001; Brenner and Wachsmuth 2012; Brenner 2013; Brenner and Schmid 2013; Curtis 2014, 2016; Florida 2017). In this respect, the globalization of economic production and consumption entails processes of economic and political restructuring, in which cities have become increasingly dependent upon markets and corporations for the provision of urban services and the creation of urban form (Castells 1996; Amen et al. 2011; Brenner 2013; Brenner and Schmid 2013; Curtis 2014, 2016; Florida 2017).

Empirically, a number of observations have been made about this process. One is a shift away from "traditional" manufacturing and services into advanced producer and financial services (Castells 1996; Friedmann 1986; Sassen 2001). A second is the liberalization of national policies on economic development, social protection and trade (Amen et al. 2011; Brenner 2013). A third is a new international division of labour, in which regional city centres have become increasingly specialized towards global production processes and services in banking (London, New York), back office outsourcing (Bangalore), electronics (Shanghai, Shenzhen) and textiles (Dhaka, Ho Chi Minh City)

(Roy 2009, 2010, 2011; Sassen 2001; Friedmann 1986; Florida 2017). A fourth is a process of uneven urban development, in which processes of speculation, inflation and dispossession have displaced marginal populations from hitherto low-value areas surrounding wetlands, water-bodies and peri-urban fringes (Harvey 2006; Johnson and Chakravarty 2013; Satterthwaite et al. 2010; Roy 2009, 2010; Webster 2011; Leaf 2011; Florida 2017).

Historically, explanations for the globalization of urban economic production and consumption are rooted in a number of material and epistemic transformations that have important bearing on the international political economy of city power. One is the collapse of the Bretton Woods system, which foreshadowed the re-conceptualization of state (and non-state) power in international politics and the growing role of markets and corporations in the planning and delivery of urban services (Brenner 2013; Brenner and Schmid 2013; Curtis 2014). A second is the revolution and ensuing globalization in information and communication technology, which facilitated the rise of transnational networks, increasingly "global" cities, and post-Fordist production regimes (Harvey 1990; Castells 1996; Sassen 2001; Brenner and Wachsmuth 2012). A third is the re-scaling of global production chains, reflecting the rise of the NICs, the BRICs and other emerging economies (Sassen 2001; Brenner 2013; Curtis 2014). A fourth and critical factor is the emergence of a scientific (and arguably wider cultural) consensus that human emissions of greenhouse gases have become the primary determinant of global environmental change, introducing a new epoch of the Anthropocene (Rockstrom et al. 2009; Sassen and Dotan 2011).

Taken together, all of these factors suggest a fundamental reassessment—or re-assertion—of the ways in which we think about cities in global politics (Brenner 2013; Brenner and Schmid 2013; Curtis 2014, 2016; Tavares 2016; Herrschel and Neuman 2017). On the one hand, cities appear to have entered a void created by the lack of national and multilateral action on climate change (Acuto 2013; Johnson et al. 2015; Acuto and Rayner 2016; Gordon and Johnson 2017). On the other, uncertainties about the ability of cities (and the states in which cities are embedded) to "manage" the complexities of globalization (and global processes like climate change) suggest a wider "crisis of statism" (Curtis 2014) that questions the legitimacy and future viability of urban planning and state-led development. In its place, privatization and decentralization of public services have been promoted (by the World Bank among others) as alternatives to the traditional planning model (Sanyal et al. 2012; Qadeer 2012; Brenner 2013; Brenner and Schmid 2013; Curtis 2014, 2016; Davidson and Gleeson 2015).

Particularly important in this regard is the seemingly contradictory observation that cities are now assuming (or being asked to assume) a leading role at precisely the time that their ability to govern has become increasingly dependent on the financialization of capital markets, the privatization of public services and the decline of the nation-state (Brenner and Wachsmuth 2012; Brenner 2013; Brenner and Schmid 2013; Curtis 2014). To take but one example (that we explore further in Chaps. 3, 4), increasing attention is now being paid to the role of management consultants and multinational corporations (such as Siemens and ARUP) in developing and designing urban mitigation and adaptation plans and policies, particularly in the Global South (McCann 2011; Bouteligier 2015; Davidson and Gleeson 2015).

Analytically, the recognition that city power is contingent upon the power of corporations, venture capital and foreign investment draws our attention to the productive forces that are now driving the political economy of labour, capital and accumulation at the urban scale (Brenner and Wachsmuth 2012; Brenner 2013; Brenner and Schmid 2013; Curtis 2014; Lee 2015; Florida 2017). In the context of climate change, it also raises questions about the extent to which cities are able to act effectively or autonomously in the absence of private actors, such as Bloomberg, the Clinton Foundation and the World Bank (Bouteligier 2015; Gordon and Acuto 2015; Acuto and Rayner 2016). Finally, it raises questions about the extent to which urban climate plans and policies are being designed to meet the largely economic and financial interests of corporations, homeowners and managerial elites, as opposed to building the resilience of slum dwellers, migrant labour and the urban poor (Boyd et al. 2015; Gore 2015; Broto Castan 2017).

2.5 Performative Power: The Ontology of Urban Form

A fourth and final way of conceptualizing the power of cities in global climate politics is to look at the ways in which cities normalize and institutionalize themselves in relation to multinational corporations, transnational networks and international regimes. One of the most striking aspects of urban climate governance is the speed and scale at which cities like London, Dhaka, Toronto and New York have integrated and embedded themselves within transnational city-networks, such as the C40, and the UNFCCC (Acuto 2013; Gordon and Acuto 2015; Acuto and Rayner 2016; Gordon and Johnson 2017). Understanding and explaining the significance of these transitions, I argue, entail looking at the Westphalian, network and corporate faces of power. But it also entails looking at the particular metrics, standards and methodologies through which cities subject themselves to different forms of climate governance (Gordon and Johnson 2017).

Performative power implies that power is deployed through processes of standardization, certification, regulation and evaluation that render cities observable and comparable in relation to corporations, networks and international regimes (Gordon and Johnson 2017). Within the field of linguistics, performativity implies that language, communication and speech are themselves repositories of power that consummate meaning and action ("with this ring, I thee wed"), thereby creating resultant norms, expectations and obligations that reflect this new sense of identity (e.g. being married) (Austin 1979). Within feminist theory, performativity embodies a "reiterative power of discourse to produce the phenomena that it regulates and constrains", highlighting the ways in which identity is created, legitimated and reproduced through narrative and discourse (Butler 1993: xii).

Taken together, theories of performativity underscore the idea that language, communication, standardization and speech are themselves forms of power that render cities observable and comparable in different governance arrangements, networks and regimes. At the same time, they call into question the ontological terms on which cities are classified for the purposes of comparison and evaluation.

Contradicting the idea that cities can be measured or understood on the basis of linear process and events (such as urban population or urban population growth), critical urbanists have argued that cities embody a metabolism that reaches out to areas and phenomena that are previously or comparatively described as "non-urban" (Brenner 2013; Brenner and Schmid 2013; Curtis 2014; Harvey 2006; cf. Lefebvre 1968 [2006]). Although the methodological contours of what we might call "a city" are by no means crystal clear (see below), the resultant conceptualization helps to capture the wider, globalizing nature of what Henri Lefebvre once called "urban society" (Lefebvre 1968 [2006]).

Underlying this perspective is both a rejection of the urban/rural dichotomy and an appreciation of the *longue duree* of systemic processes and events that connect cities and urban systems of production, consumption and exchange with ever widening labour, resources and environmental sinks (Seto et al. 2012). To quote Erik Swyngedouw and Nikolas Heynen (2003: 899):

> The socioecological footprint of the city has become global. There is no longer an outside or a limit to the city, and the urban process harbours social and ecological processes that are embedded in dense and multilayered networks of local, regional, national and global connections.

In the context of climate change, the imprecise nature of urban classification systems creates its own set of challenges for defining and measuring the ways in which cities are mitigating or contributing to climate change. Recent discussions about the challenge of measuring urban GHG emissions have debated the merits of defining emissions strictly on the basis of goods and services that are produced within a city's physical and institutional "limits" as opposed to measuring ones that are "embodied" in the goods and services that are consumed within the city but produced outside of its boundaries (Peters 2008; Peters et al. 2008; World Bank 2010a; Dahal and Niemelä 2017; Erickson and Tempest 2015; Arup-C40 2016; Singh and Kennedy 2015). However, measurements of this kind raise difficult questions about what constitutes the physical and institutional "boundaries" of urban centres (cf. Sassen and Dotan 2011; Dahal and Niemelä 2017). According to the World Bank (2010a, b: 10),

> The scope of emissions included in the city GHG Standard produced by UNEP, UN-HABITAT and the World Bank includes all emissions produced within a city, major emissions from consumption within a city, and major upstream emissions that are attributable to city residents. The question about the relevant boundaries of a city has to do with the unit to measure—strict city boundaries or the metropolitan area. A metropolitan, or functional limit of the city, may be the best scale to use, especially for larger cities.

Conceptual questions can also be raised about the appropriate indicators being used to measure urban greenhouse gas emissions. Many indexes, use an aggregate indicator of total energy consumption or GHG emissions to classify and rank urban emissions. However, aggregate indicators provide at best a partial illustration of the extent to which cities are contributing to climate change, underplaying substantially the variations that exist in terms of density, connectivity and other aspects of resource efficiency. For this reason, alternative measurements use per capita readings to capture the relative contribution of urban centres to climate change. As the World Bank (2010a, b: 10) states, "Providing emission in per capita units is helpful to highlight city-boundary issues, as most policy makers and the public can relate easily to which people are being counted".

To the extent that it is governed by cadastral mapping, industrial zoning and master planning, urbanization entails a process that continuously defines and redefines the various rights, entitlements and values that individuals, communities, corporations and states can establish in relation to urban governance regimes (cf. Satterthwaite et al. 2010; Roy 2009, 2010, 2011). To the extent that it is associated with an increase in the share of a nation's "urban" population, urbanization appears relatively clear, but the transition from rural to urban is neither straightforward nor direct. Many individuals that might be classified as "urban" in household surveys, for instance, spend considerable periods of time migrating back and forth from rural areas, highlighting the informality and seasonality of urban–rural "tele-connections" (Seto et al. 2012; Satterthwaite 2010; Saunders 2010).

Thinking about the power of cities in global climate politics therefore entails addressing the (somewhat metaphysical) challenge of determining what constitutes a city and what therefore constitutes a city's agency in relation to other forms of authority, such as corporations, city-networks and nation-states. For many urbanists (e.g. Satterthwaite et al. 2010; Seto et al. 2012; Swyngedouw and Heynen 2003; Brenner 2013; Brenner and Scmid 2013), the urban–rural distinction conveys a false dichotomy that understates the connections—or in Seto et al.'s terms, the "teleconnections"—that connect urban and non-urban spaces. First, there is clearly a desire on the part of many urbanists to reject (or at very least question rather strongly) the notion that cities may be understood in isolation from the wider (planetary) flows of energy, people, goods, services and cultural practices that constitute the urban form (Satterthwaite et al. 2010; Seto et al. 2012; Swyngedouw and Heynen 2003; Brenner 2013; Brenner and Scmid 2013). Second, the recognition that cities are inherently interconnected with non-urban spaces, places and forms challenges the notion that we might understand (and therefore ascertain the agency of) cities by questioning and rendering problematic what Brenner (2013: 91) has called the "territorial boundedness of urban units". Finally, the notion that cities are constituted by multiple flows and relations (Castells 1996) opens the possibility of observing (and therefore understanding) the ways in which urban processes affect and are affected by social, economic and ecological processes across multiple scales of analysis (Sassen and Dotan 2011; Sassen 2015).

The implication here is that cities embody processes of production, consumption, concentration and expansion (Brenner 2013) that on one hand create new patterns of urban capital formation while on the other disembedding people and communities from a local or familiar sense of place (Friedmann 1986; Sassen 2001; Harvey 2006; Satterthwaite et al. 2010; Roy 2009, 2010; Webster 2011; Leaf 2011; Brenner 2013; Brenner and Schmid 2013; Florida 2017). Within the context of climate change, such processes are reflective of the concentration of populations, energy and resources, whose ability to obtain food, shelter, mobility and access to basic human services is dependent upon the creation of vulnerability and the burning of fossil fuels. They are also reflected in the feelings of helplessness and disembeddedness that are often associated with non-local processes of global environmental change (e.g. Tschakert et al. 2013).

Particularly important in this regard is what we might call the social production of place. As Castells (1996: 442) has argued, contemporary processes of urbanization and globalization entail the creation of new spatial units that enable new flows of capital, information, technology, organizational interaction, as well as "images, sounds and symbols". In his own words, "Flows define the spatial form and processes" (Castells 1996: 439) suggesting that cities are defined primarily (or increasingly) as sites or spaces of material flows. At the same time, cities are imbued with significance and meaning that are part of people's real and imagined memories, as well as their hopes and plans and dreams.

According to Castells (1996: 453), "a place is a locale whose form, function and meaning are self-contained within the confines of physical contiguity". What this suggests is that conceptual distinctions can be made between urban spaces whose "form, function and meaning" are derived primarily as a result of external/exogenous flows and processes and those that are constructed and created in relation to particular sense of meaning through the place. Whether we ascribe agency to municipalities or to urban ecological niches, the foregoing therefore raises important questions about the historical ways in which past struggles, decisions, changes and "ruptures" affect contemporary development pathways.

2.6 Concluding Remarks: The Power of Cities in Global Climate Politics

So what does it mean to say that cities are "acting" or exerting power in global climate politics?

This chapter has explored four possibilities. The first is that city power is framed primarily in relation to the formal rules, norms and expectations that are created by states in the context international regimes, in this case, the UNFCCC. A second stems from the constellation of norms, knowledge, ideas and resources that manifest themselves in transnational city-networks. A third lies in the ability of cities to accumulate and attract the labour, resources and capital that enable them to act and exert power at a global scale. A fourth and final theory of power suggests that cities are not necessarily actors but rather, sites or spaces in which disruptions and transformations may be expected to occur.

The following chapter uses these four models of city power to frame and understand the global politics of urban climate mitigation.

References

Acuto, Michele. 2013. *Global cities, governance and diplomacy: The urban link*. New York and London: Routledge.

Acuto, M., and S. Rayner. 2016. City networks: Breaking gridlocks or forging (new) lock-ins? *International Affairs* 92 (5): 1147–1166.

Amen, Mark, Noah J. Toly, Patricia McCarney, and Klaus Segbers (eds.). 2011. *Cities and global governance: New sites for international relations*. London: Ashgate.

Anderton, K., and J. Setzer. 2017. Subnational climate entrepreneurship: Innovative climate action in California and São Paulo. *Regional Environmental Change*. doi:10.1007/s10113-017-1160-2.

Arup-C40. 2016. *Deadline 2020: How cities will get the job done*. http://www.c40.org/researches/deadline-2020. Last accessed 14 June 2017.

Austin, J.L. 1979. Performative utterances. In *Philosophical papers*, 3rd ed, ed. J.O. Urmson, and G.J. Warnock, 233–252. Oxford: Oxford University Press.

Aylett, Alex. 2015. Relational agency and the local governance of climate change: International trends and an American exemplar. In *The urban climate challenge: Rethinking the role of cities in the global climate regime*, ed. Craig Johnson, Noah Toly, and Heike Schroeder, 156–180. New York and London: Routledge Press.

Bernstein, S.F. 2001. *The compromise of liberal environmentalism*. Columbia: Columbia University Press.

Bernstein, Stephen, and Benjamin Cashore. 2012. Complex global governance and domestic policies: Four pathways of influence. *International Affairs* 88 (3): 585–604.

Betsill, M., N.K. Dubash, M. Paterson, et al. 2015. Building productive links between the UNFCCC and the broader global climate governance landscape. *Global Environmental Politics* 15 (2): 1–10.

Bouteligier, Sofie. 2015. Multinational companies and urban climate governance: Market making or successful policy innovation? In *The urban climate challenge: Rethinking the role of cities in the global climate regime*, ed. Craig Johnson, Noah Toly, and Heike Schroeder, 82–100. New York and London: Routledge Press.

Boyd, Emily, Aditya Ghosh, and Max Boykoff. 2015. Climate change adaptation in Mumbai, India. In *The urban climate challenge: Rethinking the role of cities in the global climate regime*, ed. Craig Johnson, Noah Toly, and Heike Schroeder, 139–155. New York and London: Routledge Press.

Brenner, Neil. 2013. Theses on urbanization. *Public Culture* 25 (1): 85–114. doi:10.1215/08992363-1890477.

Brenner, Neil, and Christian Scmid. 2013. The 'urban age' in question. *International Journal of Urban and Regional Research*. doi:10.1111/1468-2427.12115.

Brenner, Neil, and David Wachsmuth. 2012. Territorial competitiveness: Lineages, practices, ideologies. In *Planning ideas the matter: Livability, territoriality, governance, and reflective practice*, ed. Bishwapriya Sanyal, Lawrence J. Vale, and Cristina D. Rosan, 179–206. Cambridge: MIT Press.

Bulkeley, H. 2010. Cities and the governing of climate change. *Annual Review of Environment and Resources* 35: 229–253.

Bulkeley, H., and V. Castan Broto. 2012. A survey of urban climate change experiments in 100 cities. *Global Environmental Change* 23 (1): 92–102.

Bulkeley, Harriet, Vanesa Castan Broto, and Gareth A.S. Edwards. 2015. *An urban politics of climate change*. London and New York: Routledge.

Burch, Sarah, Alison Shaw, Freya Kristensen, John Robinson, and Ann Dale. 2015. Urban climate governance through a sustainability lens: Exploring the integration of adaptation and mitigation in four British Columbian cities. In *The urban climate challenge: Rethinking the role of cities in the global climate regime*, ed. Craig Johnson, Noah Toly, and Heike Schroeder, 119–138. New York and London: Routledge Press.

Butler, Judith. 1993. *Bodies that matter: On the discursive limits of "Sex"*. New York and London: Routledge.

Carmin, J., I. Anguelovski, and D. Roberts. 2012. Urban climate adaptation in the Global South: Planning in an emerging policy domain. *Journal of Planning Education and Research* 32 (1): 18–32.

Castan Broto, V. 2017. Urban governance and the politics of climate change. *World Development*. http://dx.doi.org/10.1016/j.worlddev.2016.12.031.

Castells, M. 1996. *The rise of the network society*. Oxford: Blackwell.
Chan, S., H. Asselt, T. Hale, K.W. Abbott, M. Beisheim, M. Hoffmann, and P. Pauw. 2015. Reinvigorating international climate policy: A comprehensive framework for effective nonstate action. *Global Policy* 6 (4): 466–473.
Chu, E., I. Anguelovski, and J. Carmin. 2015. Inclusive approaches to urban climate adaptation planning and implementation in the Global South. *Climate Policy* 16 (3): 1–21.
Curtis, S. 2014. The meaning of global cities: Rethinking the relationship between cities, states and international order. In *The power of cities in international relations*, ed. S. Curtis, 16–31. New York and London: Routledge.
Curtis, S. 2016. Cities and global governance: State failure or a new global order? *Millennium* 44 (3): 455–477.
Dahal, Karna, and Jari Niemala. 2017. Cities' Greenhouse Gas Accounting Methods: A Study of Helsinki, Stockholm, and Copenhagen. *Climate* 5 (31): 1–14.
Davidson, K., and B. Gleeson. 2015. Interrogating urban climate leadership: Toward a political ecology of the C40 network. *Global Environmental Politics* 15 (4): 21–38.
Erickson, P., and K. Tempest. 2015. Keeping cities green: Avoiding carbon lock-in due to urban development, SEI Working Paper No. 2015–11. *Stockholm Environment Institute: Seattle, WA, US*. https://www.sei-international.org/mediamanager/documents/Publications/Climate/SEI-WP-2015-11-C40-Cities-carbon-lock-in.pdf. Last accessed 14 June 2017.
Finnemore, M., and K. Sikkink. 1998. International norm dynamics and political change. *International Organization* 52 (4): 887–917.
Florida, R. 2017. *The new urban crisis: How our cities are increasing inequality, deepening segregation, and failing the middle class—And what we can do about it*. New York: Basic Books.
Friedmann, J. 1986. The world city hypothesis. *Development and Change* 17: 69–83.
Funfgeld, H. 2015. Facilitating local climate change adaptation through transnational municipal networks. *Current Opinion in Sustainability* 12: 67–73. https://doi.org/10.1016/j.cosust.2014.10.011.
Gordon, D.J. 2013. Between local innovation and global impact: Cities, networks, and the governance of climate change. *Canadian Foreign Policy Journal* 19 (3): 288–307. doi:10.1080/11926422.2013.844186.
Gordon, D.J. 2015. Lament for a network? Cities and networked climate governance in Canada. *Environment and Planning C* 34 (3): 529–545. https://doi.org/10.1177/0263774X15614675.

Gordon, D.J. 2016. The politics of accountability in networked urban climate governance. *Global Environmental Politics* 16 (2): 82–100.
Gordon, D., and C. Johnson. 2017. The orchestration of global urban climate governance: Conducting power in the post-Paris climate regime. *Environmental Politics* 26: 694–714.
Gordon, David, and Michele Acuto. 2015. If cities are the solution, what are the problems? The promise and perils of urban climate leadership. In *The urban climate challenge: Rethinking the role of cities in the global climate regime*, ed. Craig Johnson, Noah Toly, and Heike Schroeder, 63–81. New York and London: Routledge Press.
Gore, Christopher. 2015. Climate change adaptation and African cities: Understanding the impact of government and governance on future action. In *The urban climate challenge: Rethinking the role of cities in the global climate regime*, ed. Craig Johnson, Noah Toly, and Heike Schroeder, 205–226. New York and London: Routledge Press.
Hale, T. 2016. "All Hands on Deck": The Paris agreement and nonstate climate action. *Global Environmental Politics* 16 (3): 12–22.
Hale, T., and C. Roger. 2014. Orchestration and transnational climate governance. *The Review of International Organizations* 9 (1): 59–82.
Harvey, D. 1990. *The condition of postmodernity: An enquiry into the origins of social change*. Malden, MA: Blackwell.
Harvey, D. 2006. Neo-Liberalism as creative destruction. *Geografiska Annaler: Series B, Human Geography* 88 (2): 145–158.
Herrschel, T., and P. Newman. 2017. *Cities as international actors: Urban and regional governance beyond the nation state.* London: Palgrave/Macmillan.
Hickmann, T. 2016. *Rethinking authority in global climate governance: How transnational climate initiatives relate to the international climate regime.* London and New York: Routledge Press.
Hoffmann, Matthew J. 2011. *Climate governance at the crossroads experimenting with a global response after Kyoto.* Oxford: Oxford University Press.
Hughes, Sarah, and Patricia Romero-Lankao. 2014. Science and institution building in urban climate-change policymaking. *Environmental Politics* 23 (6): 1023–1042. http://dx.doi.org/10.1080/09644016.2014.921459.
Johnson, Craig, and Arpana Chakravarty. 2013. Re-Thinking the role of compensation in urban land acquisition: Empirical evidence from South Asia. *Land* 2: 278–303. doi:10.3390/land2020278.
Johnson, Craig, Noah Toly, and Heike Schroeder. 2015. Urban resilience, low carbon governance and the global climate regime. In *The urban climate*

challenge: Rethinking the role of cities in the global climate regime, 3–23. New York and London: Routledge Press.
Keohane, Robert, and David Victor. 2011. The regime complex for climate change. *Perspective on Politics* 9: 7–23.
Keohane, Robert, and David Victor. 2016. Cooperation and discord in global climate policy. *Nature Climate Change*. 9 May 2016. doi:10.1038/NCLIMATE2937.
Krasner, S. 1982. Structural causes and regime consequences: Regimes as intervening variables. *International Organization* 36 (2): 185–205.
Leaf, M. 2011. Periurban Asia: A commentary on 'becoming urban'. *Pacific Affairs* 84 (3): 525–534.
Lee, Taedong. 2015. *Global cities and climate change: The translocal relations of environmental governance*. New York: Routledge.
Lefebvre, Henri. (1968 [2006]). The urban revolution. In *The global cities reader*, ed. N. Brenner and R. Keil, 407–413. New York and London: Routledge.
McCann, E. 2011. Urban policy mobilities and global circuits of knowledge: Toward a research agenda. *Annals of the Association of American Geographers* 101 (1): 107–130.
Paterson, M., M. Hoffman, M. Betsill, and S. Bernstein. 2014. Micro foundations of policy diffusion toward complex global governance: An analysis of the transnational carbon emission trading network. *Comparative Political Studies* 47 (3): 420–449.
Peters, G.P. 2008. From production-based to consumption-based national emission inventories. *Ecological Economics* 65 (1): 13–23. doi:10.1016/j.ecolecon.2007.10.014.
Peters, G.P., and Edgar G. Hertwich. 2008. Post-Kyoto greenhouse gas inventories: Production versus consumption. *Climatic Change* 86: 51–66. doi:10.1007/s10584-007-9280-1.
Qadeer, Mohammad A. 2012. Urban development. In *Planning ideas the matter: Livability, territoriality, governance, and reflective practice*, ed. Bishwapriya Sanyal, Lawrence J. Vale, and Cristina D. Rosan, 207–232. Cambridge: MIT Press.
Rockstrom, J. et al. 2009. A safe operating space for humanity. *Nature*, September 24, 461: 472–475.
Romero-Lankao, Patricia et al. 2015. Multilevel governance and institutional capacity for climate change responses in Latin American cities. In *The urban climate challenge: Rethinking the role of cities in the global climate regime*, ed.

Craig Johnson, Noah Toly, and Heike Schroeder, 181–204. New York and London: Routledge Press.
Roy, Ananya. 2009. Why India cannot plan its cities: Informality, insurgence and the Idiom of urbanization. *Planning Theory* 8 (1): 76–87.
Roy, Ananya. 2010. Re-forming the megacity: Calcutta and the rural-urban interface. In *Megacities: Urban form, governance and sustainability*, ed. A. Sorensen and J. Okata, 93–109. London: Springer.
Roy, Ananya. 2011. The blockade of a world-class city: Dialectical images of Indian urbanism. In *Worlding Cities: Asian Experiments and the Art of Being Global*, ed. Ananya Roy and Aihwa Ong, 259–279. First Edition Blackwell Publishing.
Sanyal, Bishwapriya, Lawrence J. Vale, and Cristina D. Rosan. 2012. Four planning conversations. In *Planning ideas the matter: Livability, territoriality, governance, and reflective practice*, ed. Bishwapriya Sanyal, Lawrence J. Vale, and Cristina D. Rosan, 1–32. Cambridge: MIT Press.
Sassen, S. 2001. *The global city*, 2nd ed. Princeton: Princeton University Press.
Sassen, S. 2015. Bringing cities into the global climate framework. In *The urban climate challenge: Rethinking the role of cities in the global climate regime*, ed. Craig Johnson, Noah Toly, and Heike Schroeder, 24–36. New York and London: Routledge Press.
Sassen, S., and N. Dotan. 2011. Delegating, not returning, to the biosphere: How to use the multi-scalar and ecological properties of cities. *Global Environmental Change* 21: 823–834.
Satterthwaite, D., G. McGranahan, and C. Tacoli. 2010. Urbanization and its implications for food and farming. *Philosophical Transactions of the Royal Society of London B: Biological Sciences* 365: 2809–2820.
Saunders, D. 2010. *Arrival city: The final migration and our next world.* Toronto: Knopf Canada.
Seto, K., A. Reenberg, C.G. Boone, M. Fragkias, D. Haase, T. Langanke, P. Marcotullio, D.K. Munroe, B. Olah, and D. Simon. 2012. Urban land teleconnections and sustainability. *PNAS* 109 (20): 7687–7692.
Setzer, Joana, Laura Valente de Macedo, and Fernando Rei. 2015. Combining local and transnational action in the adoption and implementation of climate policies in the city of Sao Paulo. In *The urban climate challenge: Rethinking the role of cities in the global climate regime*, ed. Craig Johnson, Noah Toly, and Heike Schroeder, 101–118. New York and London: Routledge Press.
Shi, Linda. 2016. Adapting to climate change in cities may require a major rethink. *New Security Beat*, a blog of the Wilson Center's Environmental Change and Security Program. https://www.newsecuritybeat.org/2016/02/

adapting-climate-change-cities-require-major-rethink/?mkt_tok=eyJpIj-oiWmpGaU4yRXdabVUwTTJNMiIsInQiOiJiVEVxNldRZmxkaVJxRGQrbFlQbXVOQk9CRjU3VitEU0Uyczlkajdsd0Z4NTF6dEdhXC9nUUx6S0p3cWJ6b2hRcVgwSFBlNGV2K2ZDaXRRMUFBWFRJXC9aKzJzb0NLMGhQSkVtXC9TTlc0TlwvRlZWYUZzWkJ uV3RMUzhkV2NveVFpTEYifQ%3D%3D. Last accessed 10 May 2017.

Singh, S., and C. Kennedy. 2015. Estimating future energy use and CO_2 emissions of the world's cities. *Environmental Pollution* 203: 271–278.

Swyngedouw, E., and N.C. Heynen. 2003. Urban political ecology, justice and the politics of scale. *Antipode* 35 (5): 898–918.

Tavares, R. 2016. *Paradiplomacy: cities and states as global players*. Oxford: Oxford University Press.

Toly, Noah J. 2008. Transnational municipal networks in climate politics: From global governance to global politics. *Globalizations* 5 (3): 341–356.

Tschakert, P., R. Tutu, and A. Alcaro. 2013. Embodied experiences of environmental and climatic changes in landscapes of everyday life in Ghana. *Emotion, Space and Society* 7: 13–25.

van der Heijden, J. 2017. *Innovations in urban climate governance: Voluntary programs for low-carbon buildings and cities*. Cambridge: Cambridge University Press.

van der Ven, H., S. Bernstein, and M. Hoffmann. 2017. Valuing the contributions of nonstate and subnational actors to climate governance. *Global Environmental Politics* 17 (1): 1–20.

Webster, D. 2011. An overdue agenda: Systematizing East Asian peri-urban research. *Pacific Affairs* 84 (4): 631–642.

Widerberg, O. 2017. The 'Black Box' problem of orchestration: how to evaluate the performance of the Lima-Paris Action Agenda. *Environmental Politics* 26 (4): 715–737. doi:10.1080/09644016.2017.1319660.

World Bank. 2010a. *Cities and climate change: An urgent agenda*. Washington, DC: World Bank. http://siteresources.worldbank.org/INTUWM/Resources/340232-1205330656272/CitiesandClimateChange.pdf. Retrieved 24 May 2017.

World Bank. 2010b. *The economics of adaptation to climate change*. Washington, DC: World Bank. https://siteresources.worldbank.org/EXTCC/Resources/EACC_FinalSynthesisReport0803_2010.pdf. Retrieved 24 May 2017.

3

Cities as Saviours? The Global Politics of Urban Climate Mitigation

Abstract This chapter explores the ways in which cities have exerted and projected power in the context of global climate mitigation. Central to the analysis is an assertion that the agency of cities in global climate politics has been framed (and in many ways constrained) by the Westphalian power of the UNFCCC. By looking at a number of recent climate mitigation initiatives, this chapter explores the extent to which city-networks like ICLEI, the C40 and the Global Covenant of Mayors have been able to develop a viable alternative to the multilateral framework. Underlying the work of many city-networks is a corporate model of climate governance that entails disseminating and adopting particular standards, metrics and methodologies, suggesting new forms of performative power.

Keywords Urban climate mitigation · Cities as non-state actors
City-networks · International negotiations · The UNFCCC
The Paris Climate Agreement

C.A. Johnson, *The Power of Cities in Global Climate Politics*, Cities and the Global Politics of the Environment, https://doi.org/10.1057/978-1-137-59469-3_3

3.1 Introduction

One narrative that has generated considerable attention among scholars, activists and politicians involved in the study and practice of global climate governance is the idea that cities and city-networks are stepping in—or more appropriately—*stepping up* where national governments and international organizations have failed to act (e.g. Tavares 2016; Barber 2013, 2017).

Underlying the case for enhancing the role of cities in global climate mitigation are a number of interrelated claims.[1] One is the idea that city engagement in global climate politics reflects the inability of national governments to fulfil their responsibilities under Kyoto and Paris. In the words of Benjamin Barber (2016):

> Although the fate of nations remains uncertain, and their capacity to address global problems dubious, there is hope for both civic democracy and global action in the right and power of cities to act together across borders in our ever more interdependent world.

A second is that cities bear responsibility for addressing the causes and impacts of climate change because they account for the majority of GHG emissions (World Bank 2010a, b). As Los Angeles Eric Garcetti mayor remarked at the end of COP21,

> Cities generate 80% of the world's GDP, produce 70% of the world's greenhouse gas emissions, and house more than 50% of the world's population. The agreement's ultimate success will depend on local leadership.[2]

A third is that cities are able to "move more nimbly than national governments" (Spector 2015), legislating and ensuring the effective

[1]Following the IPCC definition, mitigation implies "a country's ability to reduce *anthropogenic greenhouse gas emissions* or to enhance natural *sinks, where ability refers to skills, competencies*, fitness and proficiencies that a country has attained and depends on technology, institutions, wealth, equity, *infrastructure and information*".

[2]https://www.lamayor.org/statement-los-angeles-mayor-eric-garcetti-and-33-other-us-climatemayors-adoption-historic-paris. Last accessed 7 December 2016.

implementation of actions aimed at reducing urban GHG emissions (cf. Bulkeley et al. 2015). Spector (2015), for instance, argues that the Compact of Mayors:

> ...essentially parallels the national process enshrined in the COP21 treaty, but at the local level. It also serves as a hedge against recalcitrant national governments: if they backslide, or a new party takes power and rejects the climate agreement, a decentralized network of cities will keep the process going.

A fourth and related argument is that cities are more familiar with local needs, issues, complexities and interests, and they are therefore uniquely placed to understand and address the challenge of reducing emissions at the urban scale. In the words of Zenghelis and Stern (2016):

> Implementing policy action to reduce emissions is often easiest at the city level. This means increased effort is required to devolve administration to the urban level and empower cities through greater fiscal autonomy. Things like recycling schemes, energy from waste, broadband networks, integrated public transport systems, "smart" buildings and biking networks are easy to appreciate and can come with popular local support.

All of these are compelling arguments that merit careful attention. However, asserting that cities *should* play a role in governing climate change does not imply that they will, highlighting the ways in which climate change norms and objectives are framed, defended and adopted at the urban scale (Bulkeley and Betsill 2013; Bulkeley et al. 2015; Castan Broto 2017). Although cities are clearly (re)positioning themselves as critical actors in the global fight against climate change, their ability to "act"—i.e. to legislate, regulate and implement policies on climate change—remains highly dependent upon the constitutional, regulatory, legislative and financial powers that they have at their disposal and their ability to reconcile the real and perceived opportunity costs of addressing climate change at the expense of other more pressing urban priorities (Bulkeley and Betsill 2013; Bulkeley et al. 2015; Sassen and Dotan 2011; Bouteligier 2015; de Flander 2015; Castan Broto 2017; van der Heijden 2017).

Complicating and potentially undermining the notion that cities are somehow uniquely placed to effect change on climate change is the notion that cities are also embedded in a complex web of social, political and financial relations that can undermine and overwhelm their ability to act effectively and collectively. One relatively hard constraint that has important bearing on the power of cities in global climate politics is the broad constellation of powers, responsibilities and requirements that are conferred—and often imposed—onto cities by nation-states. Although city engagement in global climate politics is arguably an effort on the part of cities to circumvent these limitations, their ability to do so is framed and constrained by the power and political preferences of nation-states. As Davis (2016) observes, "it would be naive to think that national governments will sit by idly as cities and mayors seek to enter and guide global conversations on their own terms".

Recent empirical work on urban and polycentric climate governance has shown that cities can develop innovative ways of reducing emissions when they are "nested" within national political systems that specify the rights of local municipal bodies to enact plans and policies governing climate-sensitive sectors in energy, transportation and infrastructure (Pahl-Wostl 2009; Ostrom 2010; Johnson et al. 2015). There is also evidence that national governments can "nudge" municipalities into action by providing financial, regulatory and ideational/political support for local climate initiatives (Tanner et al. 2009; Ostrom 2010; Aylett 2015; Gore 2015; Johnson et al. 2015). However, decentralization also entails processes of "orchestration", standardization and surveillance that effectively subject cities to new forms of obligation, financial reporting and control (Hale and Roger 2014; Chan et al. 2015; Abbott et al. 2016; Gordon and Johnson 2017). The implication here is that cities will find it difficult to adopt ambitious climate change commitments that deviate from national policy norms and objectives.

A second and related constraint stems from the financial dimensions of acting unilaterally or collaboratively on climate change. In the case of mitigation and adaptation (the latter of which we address in Chap. 4), acting on climate change entails an ability to invest in actions that typically exceed the revenue base of many local authorities. As we have already noted in Chap. 2, cities have become increasingly dependent upon markets, corporations and external aid transfers (from bilateral and multilateral donors, as well as from higher levels of government) for financing the long-term capital investments that are necessary for reducing emissions in high-carbon sectors, such as transportation, energy and housing. Consequently, multinational corporations, such as Siemens and Arup, have become increasingly engaged in developing plans and policies that cities are now using to tackle climate change (Arup-C40 2016; Bouteligier 2015; Gordon and Acuto 2015).

A third factor that potentially tempers the power that is being ascribed to the role of cities in global climate politics is what we might call the messy and unwieldy nature of urban politics. As noted in Chap. 2, cities are often seen as important sites of innovation, where new and experimental approaches to climate governance have been introduced and disseminated in other policy settings (Bulkeley and Castan Broto 2012; Bulkeley et al. 2015; Anderton and Setzer 2017). However, cities are also complex agglomerations of people, finances, labour and capital that are at best only loosely connected to local efforts at mitigating and adapting to climate change (Davis 2016; Castan Broto 2017). To suggest that formal commitments on the part of city leaders will be met with a "groundswell" of support for public action on climate change may overstate the power of urban policy and popular support for climate policy (Acuto 2013; Davis 2016; Castan Broto 2017).

Fourth, the evidence that has been used to document city, non- and sub-state action on climate change can be challenged for overstating the impact of local climate policy initiatives on reducing emissions (Widerberg and Stripple 2016; van der Heijden 2017), suggesting that the formal policy responses of cities and municipalities may have limited bearing on the ways in which people in cities actually go about acquiring, consuming and disposing of waste and resources. In Diane Davis' words (2016):

> Even if nation-states do devolve significant autonomy to their cities, it is not clear that local governance will be a panacea. Problems with indefinite borders, after all, pose challenges to any established borders of governance — including cities. While local governments may offer more space for experimentation in governance styles and strategies by virtue of their size and number, there is no guarantee that cities will be more successful or innovative.

Understanding the power of cities in global climate politics therefore entails identifying the core constituencies and networks of resistance and support for building low-carbon, resilient cities (Brenner 2013; Brenner and Schmid 2013; Geels 2004; Geels and Raven 2006; Geels and Schot 2007; Kennedy 2011; Bulkeley et al. 2015). Whether cities emulate or engage in climate policy networks depends on the ways in which politicians, officials, epistemic and other policy communities frame and define policy discourses, as well as the rules, norms and expectations that are used to justify local actions, commitments and decisions (Bouteligier 2015; Bulkeley et al. 2015; Funfgeld 2015; Gordon and Acuto 2015; Lee 2015; McCann 2011).

A final issue that has critical bearing on our understanding of city power in global climate politics is the broader impact of city actions on state, sub- and non-state actors in global climate politics. As noted in the previous chapter, the UNFCCC is often characterized as a weak regime that lacks the capacity to force compliance on the part of its member states (Keohane and Victor 2011, 2016). For this reason, the power that is now being ascribed to the UNFCCC appears to be more squarely rooted in the diffusion of policy norms, standards and commitments that may be used in framing, legitimating and orchestrating new forms of climate action (Hale and Roger 2014; Paterson et al. 2014; Betsill et al. 2015; Faulkner 2016; Gordon and Johnson 2017). Under the auspices of the Paris Climate Agreement, the expectation is that member states (along with sub- and non-state actors) will develop their own strategies for maintaining global temperatures at 2 °C above pre-industrial levels, and ultimately pursuing the aspirational goal of stabilizing them at 1.5 °C (see below). For cities (and city-networks), the Agreement offers new opportunities for reducing the urban (and therefore global) carbon footprint. In December 2016, for instance, the C40 and the multinational consulting firm ARUP announced a detailed plan for achieving the 1.5 °C target (see below).

Whether these and other plans and commitments will lead to substantive, transformative changes on the ground—or, more appropriately, in the streets—will depend on a number of factors, including the ability of national and sub-national (including urban) policies to affect long-term behavioural change and the ways in which local mitigation efforts affect actions and decisions at wider scales of interaction.

On the first point, climate policies have been criticized for adopting standards and commitments that have very little bearing on people's (including corporations' and government's) actions and decisions (cf. Widerberg and Stripple 2016). On the second, there is the larger free rider problem that successful efforts to reduce emissions may actually reduce the incentive of other actors (including nation-states) to take action themselves (cf. Hsu et al. 2015). Framed positively, early and ambitious action on climate change may provide a model on which other jurisdictions may frame and legitimate ambitious climate policy actions of their own. However, the opposite may also be true—i.e. that pioneering efforts of local authorities will be ineffectual (and therefore they may lose their support) if they reduce or remove the incentive for other actors (including cities and nation-states) to act on climate change.

All of which raises important questions about the extent to which and ways in which climate change appears on local policy agendas, and how these long-range, transformational norms and objectives are framed, defended and adopted at the urban scale (Bulkeley and Betsill 2013; Bulkeley et al. 2015; Aylett 2015; Anderton and Setzer 2017). Because of their long-time horizons, climate change commitments are often low on the policy agenda of local politicians, officials and other stakeholders, reflecting the real and perceived opportunity costs of reducing urban emissions at the expense of other more pressing urban priorities (Sassen and Dotan 2011; Bouteligier 2015; de Flander 2015; Keohane and Victor 2011, 2016; Aykut 2016). Although policy analysts are now calling upon cities to mainstream climate change into operational planning and policy (e.g. World Bank 2010a, b), the reality is that many cities lack the constitutional, regulatory, legislative and financial power to act effectively and autonomously on climate change, highlighting the economic and political disparities that separate cities and nation-states (Gore 2015; Johnson et al. 2015; van der Heijden 2017).

This chapter explores the implications of this apparent transformation, reflecting critically on the ways in which city engagement in global climate politics is affecting our understanding of city power within the global system. The first form of power is framed primarily in relation to the formal rules, norms and expectations that are created by states within the context of the UNFCCC.

3.2 Westphalian Power: City Engagement in the Multilateral Regime

With the benefit of hindsight, the Kyoto Protocol now seems a modest attempt at coordinating international action on climate change. The policy critiques are well-known (Hoffmann 2011; Keohane and Victor 2011, 2016; Paterson et al. 2014; Bulkeley et al. 2014): the Protocol lacked a credible compliance mechanism for verifying emission reductions; it failed to incorporate some of the world's largest emitters (e.g. China and the USA); it was arguably too dependent upon economic instruments, such as carbon offsets and emissions trading schemes; it provided little in the way of stable and secure financing for adaptation and mitigation efforts in the Global South (cf. Biermann 2011; Keohane and Victor 2011, 2016; Hurrell and Sengupta 2012; Hochstetler and Milkoreit 2013; Stevenson and Dryzek 2014).

Prior to Paris, sub- and non-state actors (including cities, provinces and local communities) were able to use the failure of Kyoto as a way of justifying more ambitious commitments at the local and urban scale (Aylett 2015; Burch et al. 2015; Johnson et al. 2015; Setzer et al. 2015). Making specific reference to Kyoto and the UNFCCC, many cities were able to successfully frame and legitimate new forms of urban climate policy (Bulkeley and Betsill 2013), and at the same time enhance their political profile on the international stage (Acuto 2013; Gordon and Acuto 2015).

During the run-up to COP21, international negotiators were particularly concerned with addressing these and other shortcomings by pursuing the following objectives:

- Reducing global emissions to maintain global temperatures at 2 °C above pre-industrial levels (and ultimately pursuing an aspirational goal of 1.5 °C);
- Achieving emission reductions commitments on the part of "non-Annex 1" countries, such as China, India, Indonesia and Brazil;
- Providing stable financing for adaptation and mitigation, primarily in the form of the $100bn per year commitment to the Green Climate Fund;

- And in what was widely perceived to be a concession to the small island states, making progress on the so-called Loss and Damage mechanism that came out of the Cancun Adaptation Framework.[3]

At the heart of the Paris Climate Agreement are the "Intended Nationally Determined Contributions"—or INDCs—that member states had already developed in advance of the meetings at COP21. Here, the expectation was that the INDCs would be used (or revised in the form of Nationally Determined Contributions) as the basis for reducing global emissions to remain within the 2 °C target. National progress on the NDCs will be subject to a third-party peer review that will take place every five years after the ratification of the agreement. Notably, Article 118 of the Agreement "welcomes the efforts of non-Party stakeholders (including cities) to scale up their climate actions, and encourages the registration of those actions in the Non-State Actor Zone for Climate Action", a point I take up below.

The 2015 Paris Climate Agreement therefore embodies a conscious effort to move away from time-bound targets and deadlines to a more flexible "pledge and review" mechanism that outlines new reporting and compliance mechanisms for nation-states (Falkner 2016; Keohane and Victor 2016). From a geopolitical perspective, such flexibility was instrumental in bringing otherwise reluctant partners (e.g. China, India, and the USA, ca. 2015) into a global agreement. From an environmental policy perspective, it left much to be desired. According to one report that was released by the UNFCCC Secretariat shortly before COP21,[4] a successful agreement is projected to slow the growth in global emissions of carbon dioxide by about 4 billion tons per year by the year 2030. However, without further action, it estimates that global average temperatures will rise by 2.7 °C by the year 2100. Another recent study by the World Bank suggests that already existing plans

[3]Here it's worth noting that although loss and damage received favourable mention in the final agreement, it contained the explicit proviso that recognizing or observing the existence of loss and damage "does not involve or provide a basis for any liability or compensation" (Paris Agreement, Article 52; Johnson 2017). More on this in Chap. 4.

[4]http://newsroom.unfccc.int/unfccc-newsroom/indc-synthesis-report-press-release/. Last accessed 6 May 2016.

to develop coal-fired power plants in China, India, Indonesia and Vietnam will "spell disaster" for the Paris Accord and for the planet.[5]

In principle, the Paris Agreement provides ample scope for enhancing city involvement in climate mitigation. The language is explicit, for instance, in recognizing the role that cities as "non-Party stakeholders" will play in "scaling up" their climate actions, documenting these actions through the NAZCA platform. The Agreement also recognizes "the need to strengthen knowledge, technologies, practices and efforts of local communities and indigenous peoples related to addressing and responding to climate change", highlighting the importance of financing and technology transfers in building capacity for climate adaptation and mitigation. Finally, Article V (137) of the Agreement reiterates the importance of national policies in framing the possibilities for climate action at the urban scale.

Taken together, the Paris Agreement is clearly framed in relation to the formal roles and responsibilities that are outlined explicitly in the LPAA and the UNFCCC, vesting considerable responsibility in the reporting mechanisms that national governments will put in place to evaluate the performance of their NDCs. The Agreement provides ample text on capacity-building (and related fields, such as finance and technology transfer), but developing institutional mechanisms for data collection and evaluation will present challenges. Notwithstanding the (monumental) task of ensuring compliance on the part of member states, the agreement implies considerable resources, research facilities, reporting procedures and compliance mechanisms that are arguably in short supply in low-income countries and regions. Building the capacity of national and sub-national governments to achieve and report on their progress will entail new forms of financing (see below) to ensure that the data being used to inform national emissions reporting are accurate, reliable and transparent.

Within this context, it has become increasingly apparent that cities (as well as other sub- and non-state actors) will play—or will be expected to play—a critical role in achieving the aims and aspirations

[5]http://www.theguardian.com/environment/2016/may/05/climate-change-coal-power-asia-world-bank-disaster?CMP=share_btn_tw. Last accessed 10 May 2016.

of the Paris Climate Agreement (Chan et al. 2015; Hsu et al. 2015; Hale 2016). For our purposes, the critical question is whether the new roles that are being ascribed to cities will translate into new forms of power in global politics.

3.2.1 All that States are not: Cities as Sub- and Non-state Actors

According to Hale (2016: 13), cities are part of a broader category of sub- and non-state actors that have become "more central" to the UNFCCC, creating what he has called a "new catalytic and facilitative climate regime". Particularly important in this regard is the notion that national governments and international organizations (such as the UNFCCC Secretariat) have become increasingly receptive to the idea of cities, corporations, provinces and civil society organizations deciding and advancing emission reductions actions and commitments of their own.[6] However, whether these and other sub- and non-state actors constitute new forms of power depends on the extent to which they are able to act and potentially resist the authority that nation-states are projecting through the multilateral regime.

Although still in their infancy, interpretations of the implications of sub- and non-state engagement in the post-Paris climate regime highlight a number of questions about the power of cities in relation to international organizations, multinational corporations and nation-states. First, questions have been raised about the extent to which the formal recognition of cities represents new forms of city power, as opposed to changing patterns of "orchestration" and control

[6]At the heart of Hale's argument are three inter-related observations about the events that unfolded prior to and during COP21. First, in the run-up to Paris, national governments (particularly the host governments of France and Peru) and international organizations (primarily the UNFCCC Secretariat and the UN Secretary General), "actively sought to galvanize sub/non-state climate action as a way to achieve their mitigation and adaptation goals" (Hale 2016: 13). Second, the French Presidency declared the LPAA "a fourth pillar of COP21 alongside, and equal to, the national pledges, the financing package and the negotiated agreement" (Hale 2016: 14). Finally, the inclusion of "non-Party stakeholders" in the final text recognised sub- and non-state actors "as a core element of its logic of spurring rising climate action over time" (Hale 2016: 14).

that both challenge and reify the power of nation-states (Gordon and Johnson 2017). As Widerberg and Stripple (2016: 489) have argued, "Governments are by no means passive bystanders" in this process, and the apparent rise of cities needs to be interpreted in relation to existing forms of power, including those of nation-states". According to Hale (2016: 16), a number of member states opposed earlier drafts of the LPAA precisely because it made specific reference to cities and the private sector:

> … some countries like Sudan worried that such references could undermine their sovereignty, and more ideological regimes like Venezuela (were) concerned about the prominence that would be given to multinational corporations.

A second and related question concerns the analytic and political implications of classifying cities alongside corporations, provinces, NGOs and other sub- and non-state actors. Although cities, provinces, companies and NGOs arguably share a common experience of being marginalized historically within the global climate regime, the idea that they can be classified under a single category of sub- and non-state action seems to underplay the very different interests and incentives that are motivating their involvement (or the lack thereof). As Hale (2016: 20) asks, "does the pragmatic, problem-solving approach that brings business and sub-national jurisdictions into the regime risk detracting from questions of climate justice and an equitable distribution of the atmosphere's resources?"

A third question concerns the wider implications of city involvement on the integrity of the multilateral regime. During and after the Kyoto period, observations have been made that global climate governance has become increasingly "fragmented" by the appearance of new and competing forms of authority. According to Biermann and Pattberg (2012: 13), the transformation has been "characterized",

> … by an increasing segmentation of different layers and clusters of rule making and rule implementing, fragmented vertically among supranational, international, national and subnational layers of authority

(multilevel governance) and horizontally among different parallel rule-making systems maintained by different groups of actors (multipolar governance).

Whether we associate the appearance of new governance actors, layers and mechanisms with a dilution (as opposed to an augmentation) of localized power, the argument being made here is that the rise of sub-national actors constitutes a significant departure from the historical principle of central state sovereignty and responsibility (Biermann and Pattberg 2012; Bulkeley et al. 2014; Hsu et al. 2015; Hale 2016; Widerberg and Stripple 2016). Within the context of multilateral climate governance, an overarching concern is that the growing involvement of cities as sub- or non-state actors will "enable governments to side-step their own obligations, lowering the overall ambition of mitigation and adaptation goals" (Hsu et al. 2015: 501).

Fourth, there is a deeper methodological challenge of deriving inferences from what are largely self-reporting mechanisms as a means of documenting all of the possible climate-related actions and commitments that are being made by non-state actors, including cities. By the NAZCA Website's own admission, the (NAZCA) "portal has been constructed to rely on the underlying data providers",[7] suggesting that the data being provided are prone to errors of validity, consistency and comparability across time and space (cf. Hsu et al. 2015; Widerberg and Stripple 2016). There is also the problem of determining whether national and sub-national commitments are actually having a discernible impact on climate policy outcomes (cf. Keohane and Victor 2016). As Widerberg and Stripple's analysis (2016: 493) of five emission reduction databases (including NAZCA) suggests, methodologies of this kind "lack information on the ex post impacts of initiatives, both by qualitative reporting through, for example, progress reports, or through quantitative reporting such as calculations of achieved emission reductions". Elsewhere, Hsu et al. (2016) have argued that the NAZCA platform "captures just a fraction of what is being

[7]http://climateaction.unfccc.int/about. Last accessed 11 May 2016.

done", highlighting the methodological and financial challenges of documenting mitigation activities in cities, regions and countries where the capacity for reporting and evaluating climate policy actions is weak.[8]

The implications of cities subjecting themselves to observation and self-comparison we take up below, but here it is also worth noting the methodological challenges of establishing what constitutes effectiveness and impact in the post-Paris climate regime. As Jeroen van der Heijden (2017) has recently argued, establishing the effectiveness and impact of mitigation policies entails isolating and evaluating the relative importance of "program context" and programme design. Particularly challenging is what he calls the "leadership delusion", in which "those involved in the development, administration and implementation of voluntary programmes – as well as those who voluntarily commit to them – have strong incentives to present a narrative of successful performance even where there is none" (van der Heijden 2017: 4).

Van der Heijden's comments draw our attention to the methodological challenges of interpreting the observations and declarations of actors (e.g. city mayors and politicians) who are themselves embedded in local and urban processes of climate governance.

Taken together, the empirical insights that are now being generated in the context of the Paris Agreement have raised important questions about the power of cities and sub- and non-state actors in relation to nation-states that are as yet poorly understood. Particularly important from a global politics perspective is the extent to which cities and city-networks are able to influence and offer alternatives to the multilateral regime.

By and large, the major form of city involvement is through larger city-networks, such as ICLEI, the C40 and the Compact of Mayors.

---

[8]That said, networks of international researchers are currently undertaking concerted efforts to improve the methodological standards on which voluntary climate mitigation actions and commitments are communicated and evaluated at different scales of analysis. Important recent examples of this kind include the Initiative for Climate Action Transparency (http://www.climateactiontransparency.org) and Galvanizing the Groundswell (http://www.climategroundswell.org/).

3.3 Network Power

Closely tied to the discussion of whether and to what effect cities are engaging with the UNFCCC is the notion that cities are also exerting power through transnational city-networks, such as the Cities Climate Leadership Group (C40), ICLEI (Local Governments for Sustainability) and the Global Covenant of Mayors for Climate and Energy. Given the limited extent to which national governments were able to reduce emissions during the Kyoto period, increasing attention has been paid to the role of transnational city-networks, whose membership activities have entailed policy advocacy, "paradiplomacy", awareness raising and the dissemination of norms, knowledge and resources aimed at reducing emissions at the urban scale (Toly 2008; Gordon 2013; Bouteligier 2015; Bulkeley et al. 2015; Gordon and Acuto 2015; Johnson et al. 2015; Lee 2015; Acuto and Rayner 2016; Gordon 2016; Tavares 2016; Gordon and Johnson 2017). To quote Harriet Bulkeley (2010: 233), "urban networks have taken a more overtly political stance toward the issue, seeking to position cities as critical sites for addressing the issue of climate change or even opposing national governments".

Arguably, the most widely cited examples of city-networks working on transnational climate policy issues are ICLEI, the C40 and the more recently established Global Covenant of Mayors (Bulkeley 2010; Gordon and Acuto 2015; Acuto and Rayner 2016; Hickmann 2016; Gordon and Johnson 2017). First established in 1990, ICLEI has a membership of more than 1500 cities, towns and regions,[9] offering a wide range of programmes aimed at reducing emissions at the urban scale, including the *Green Climate Cities Program* (formerly the *Cities for Climate Protection Program*), the *Urban Low Emission Development Strategies*, carbon*n*, the *Global Protocol for Community-Scale GHG Emissions Inventories* and the *Global Covenant of Mayors* (cf. Betsill and Bulkeley 2004; Hickmann 2016). The C40 was established in London in 2005, offering programmes on City Diplomacy, the Global Protocol for Community-Scale GHG Emissions Inventories and the Global

[9] http://www.iclei.org. Last accessed 9 December 2016.

Covenant of Mayors (Gordon and Acuto 2015). According to its Website,[10] the C40 has a membership of "more than 80 of the world's greatest cities, representing over 600 million people and one quarter of the global economy". Finally, the Global Covenant of Mayors was established (with the support of ICLEI and the C40) in 2016, following the merger of the Compact of Mayors and the European Covenant of Mayors. According to its Website,[11] the Global Covenant represents "more than 7100 cities from 119 countries and six continents, representing more than 600 million inhabitants, over 8% of the world's population".

As noted in Chap. 2, early instances of city-network engagement on climate change were largely aimed at promoting and raising awareness about the need to reduce emissions at the urban scale (Bulkeley 2010; Hickmann 2016; Gordon and Johnson 2017). In recent years, cities and city-networks have become increasingly involved in governing climate politics, establishing baselines, disclosure platforms and measurement protocols that many cities are now using to reduce their emissions and (critically) report their actions and commitments to other cities, city-networks and the UNFCCC (Bansard et al. 2016; Hickmann 2016; Bäckstrand and Kuyper 2017; Gordon and Johnson 2017; van der Ven et al. 2017). The Local Government Climate Roadmap, for instance, identifies three "overarching goals" of Recognition (e.g. through COP decisions), Engagement (e.g. in ADP Technical Examination Processes, NAZCA Platform, High Level Dialogues and special accreditations) and Empowerment (e.g. with new financing programmes) (ICLEI 2015: 23).

However, one of the challenges of establishing the power of transnational city-networks is that many of the networks that have been most active in urban climate governance are also working closely with (and, indeed, deriving their legitimacy from) the UNFCCC (cf. Hickmann 2016; Gordon and Acuto 2015; Gordon and Johnson 2017). Another is that many of the networks that are most active are also actively

[10] http://www.c40.org/about. Last accessed 9 December 2016.

[11] https://www.compactofmayors.org/globalcovenantofmayors/. Last accessed 9 December 2016.

involved *with one another*. Both ICLEI and the C40, for instance, have been actively involved in supporting the Global Covenant of Mayors. Both were also instrumental in launching the Global Protocol for Community-Scale GHG Emissions Inventories (Hickmann 2016), a methodology that "harmonizes greenhouse gas emissions measurement and reporting processes for cities of all sizes and geographies, and allows them to plan and finance climate action".[12] Making sense of these networks—and disentangling their actions and impacts from those of the UNFCCC and the private sector—poses a number of analytical challenges.

Theories of transnational network governance suggest that networks can play a critical role in ordering priorities, framing expectations and defining what constitutes appropriate behaviour (Finnemore and Sikkink 1998; Toly 2008; Bernstein and Cashore 2012; McCann 2011 Paterson et al. 2014; Betsill et al. 2015; Bulkeley et al. 2015; Lee 2015; Gordon and Acuto 2015; Gordon 2016; Gordon and Johnson 2017). Similarly, in the context of global climate governance, transnational city-networks appear to provide a number of functions that have important bearing on the power of cities in global climate politics.

First, they provide critical points of entry and access to finance, technology and expertise used in the development or low-carbon and climate-resilient pathways (Chu et al. 2016; Carmin et al. 2012; Gordon and Acuto 2015; Gore 2015; Hughes and Romero-Lankao 2014; Lee 2015; Funfgeld 2015; Romero-Lankao et al. 2015; Setzer et al. 2015; Acuto and Rayner 2016). An important function here is the role that transnational networks play in pooling resources, sharing ideas and building capacity at the urban scale (Toly 2008; Bulkeley 2010; Funfgeld 2015; Gordon and Acuto 2015; Lee 2015; Gordon and Johnson 2017).

Second, transnational networks facilitate the diffusion of policy norms, standards and commitments that can influence the actions of members and non-members. Through conferences, workshops and other forms of social interaction, city-networks provide important

[12]http://www.iclei.org/activities/agendas/low-carbon-city.html. Last accessed 9 December 2016.

models and metrics that may be used in framing urban mitigation plans and policies (Bouteligier 2015; Gordon 2013; Gordon and Acuto 2015; Gore 2015; Setzer et al. 2015). Cities also frequently use other cities as models or "benchmarks" for their own policy innovations (Toly 2008; Leichenko 2011; McCann 2011; Romero-Lankao and Dodman 2011; Bulkeley and Castan Broto 2012; Carmin et al. 2012; Burch et al. 2015; Chu et al. 2016; Lee 2015; Setzer et al. 2015).

Third, engaging in transnational networks—either through formal membership or informal affiliations—provides an important means of showcasing new policy initiatives, disseminating best practices and solidifying one's place within other transnational policy networks, such as the UNFCCC. At the international level, benefits of this kind can enhance the profile of cities and city leaders, thereby improving recognition, communication and access to international capital (Acuto 2013; Gordon and Acuto 2015; Gordon and Johnson 2017). At the subnational level, transnational city engagement can also enhance the profile of city leaders in the eyes of local constituents, particularly when it results in new forms of investment that produce tangible co-benefits in the form of infrastructure.

Leading, joining, following and gaining access to transnational city-networks therefore provide an important means by which cities can project and situate themselves more effectively in local and international policy arenas. However, the factors affecting transnational city engagement and the impacts of city-networks on urban climate governance remain poorly understood, reflecting the myriad ways in which cities and other urban interests may conceivably interact with global climate policy networks and processes (Bouteligier 2015; Gordon and Acuto 2015; Gore 2015; Lee 2015; Johnson et al. 2015; Romero-Lankao et al. 2015; Setzer et al. 2015). Notwithstanding the prestige that may attract leaders of large cities like London and New York (Acuto 2013), city-networks possess few mechanisms for monitoring or enforcing the actions and decisions of their members (Gordon 2013; Lee 2015). Moreover, participating in transnational forums is costly, and the resources being provided by networks like the C40 and ICLEI are modest (Lee 2015; Gordon 2013).

Recent empirical work suggests a tendency on the part of transnational networks like the C40 and the Carbon Neutral Cities Alliance to cluster around larger, wealthier cities, whose economic, political and diplomatic clout enables them to occupy pivotal niches within climate policy networks (Acuto 2013; Gordon 2013, 2016; Gordon and Acuto 2015; Lee 2015; Davis 2016; Acuto and Rayner 2016; Gordon and Johnson 2017). By virtue of their leadership in the C40 and the CCI, cities like London and New York have been able to position themselves as key players, forging new opportunities for proactive mayors like (former New York Mayor) Michael Bloomberg and (former London Mayor) Ken Livingstone to showcase their efforts in the global fight against climate change.

At the same time, cities like Lima, Addis Ababa and (until recently) Los Angeles appear to have occupied what was clearly a marginal status, suggesting variations of interest and involvement in the wider movement (Acuto 2013). Moreover, there are many wealthy cities (e.g. Miami, Manila and Berlin) that have not engaged extensively in transnational climate policy networks, and many low-income cities that have (Gordon and Acuto 2015; ICLEI 2015). According to carbon*n*, an international registry that collects information about sub-national actions and commitments on climate change, more than 50 cities in Thailand and Tanzania reported carbon reduction efforts in 2015; for Britain, the comparable figure was two (ICLEI 2015).

How do we explain these disparities? And what implications do they have for the study of urban and global climate governance? One possibility is that cities *are* taking action on climate change, but their actions are not being documented in registries like carbon*n*. A second is that cities are choosing not to share their actions with *these* particular networks, raising interesting questions about the ways in which and extent to which voluntary transnational networks are able to exert power and influence in global climate politics (Chap. 4).

Theories of city-network governance suggest that regional proximity and social familiarity in the form social capital can play an important role in facilitating flows of norms, information, resources, ideas and people that might lead to stronger forms of policy diffusion and network engagement (Lee 2015). Lee (2015), for instance, argues that network engagement (and formation) is contingent upon the temporal and socio-spatial factors that connect cities (and city leaders) with transnational (or what he calls "translocal") relations and elites. Cities, he argues, are more likely to emulate and learn from other cities that share similar geographical, regional and sociocultural attributes—what he calls "homophily" and an established record of successful policy implementation (Lee 2015: 74–78).

What this suggests is that networks have effects on members as well as non-members, facilitating the diffusion of norms that may lead to stronger forms of integration over time and space (Finnemore and Sikkink 1998; Toly 2008; Bernstein and Cashore 2012; McCann 2011; Paterson et al. 2014; Betsill et al. 2015; Bulkeley et al. 2015; Lee 2015; Gordon and Acuto 2015; Gordon 2016; Gordon and Johnson 2017). McCann (2011: 109), for instance, identifies the "socially produced and circulated forms of knowledge"—what he calls "urban policy mobilities"—that shape the content and direction of urban plans and policies, suggesting that cities are on the one hand fixed in relation to infrastructures, path dependencies and political interests and on the other subject to geographical flows of capital, people, information and ideas that make cities more or less connected to wider networks (McCann 2011). Similarly, Toly (2008: 345) makes the case that cities and city-networks can act as "norm entrepreneurs" that "mobilize support for 'particular standards of appropriateness' and convince states, potential norm leaders, to adopt these standards".

At the same time, questions have been raised about the overall effectiveness of city-networks in changing behaviour at the urban scale. Much like the Global Covenant, ICLEI employs a 5-step methodology that (1) conducts a GHG emissions inventory; (2) adopts emission reduction targets; (3) develops a local action plan; (4) implements relevant policies and measures; and (5) monitors and verifies the outcomes and results. However, evaluations of the methodology have shown that many cities have not completed the milestones (Reams et al. 2012; Hickmann 2016), and that the milestones have been used to "redefine already existing policies" as opposed to challenging or changing the status quo (Hickmann 2016: 77).

A final point that takes us into the next two sections of this chapter is that joining, emulating and even trying to join formal networks entails subjecting oneself to rules, standards and expectations that are themselves forms of power (cf. Toly 2008; Gordon and Johnson 2017). The C40, for instance, offers four types of affiliation that effectively differentiate (and stratify) cities within the network: "megacities" (including Madrid, Buenos Aires and Cape Town), "innovator cities" (such as Rotterdam, Quito and Changwan), "steering committee cities" (Paris, London, Venice, Durban, Boston, Los Angeles, Mexico City, Jakarta, Tokyo, Hong Kong and Seoul) and "observer cities". According to the C40, *megacities* are cities with a city population of 3 million or more, and/or a metropolitan area population of 10 million or more or they are one of the top 25 global cities ranked in terms of GDP. *Innovator cities* are "cities that do not qualify as megacities but have shown *clear leadership* in environmental and climate change work", including international recognition for "barrier-breaking climate work". *Observer cities* are cities that have applied to join the C40 and are currently awaiting a decision. Finally, *steering committee cities* are cities that have assumed a leading role in defining and taking forward climate policy initiatives on behalf of the network.

Insofar as transnational city-networks include cities that are not necessarily formal members, we can therefore think of city engagement on a broad spectrum that encompasses many different forms of action, decision-making and resistance, including basic acknowledgement of network norms and procedures to extensive leadership and material contributions. On this basis, useful distinctions can be made between "early adopters", cities that establish new and experimental approaches to urban climate governance, and "late adopters", cities that alter their actions and decisions on the basis of these new norms, practices and experiences (Bulkeley and Castan Broto 2012; Bulkeley et al. 2015; Carmin et al. 2012; Lee 2015). Whether cities emulate or adopt a particular policy objective depends on the ways in which politicians, officials and local and epistemic communities frame and define policy discourses, as well as the rules and procedures that determine who can legitimately take part in decision-making processes (Bouteligier 2015; Bulkeley et al. 2015; Funfgeld 2015; Gordon and Acuto 2015; Lee 2015; McCann 2011).

What this suggests is that engaging in transnational networks is a dynamic process that varies in terms of frequency, duration, intensity (measured in terms of material resources contributed and extracted by individual members) and the relative formality of rules and institutional arrangements connecting cities and transnational city-networks (Acuto 2013; Funfgeld 2015; Gordon and Acuto 2015; Lee 2015; McCann 2011).

- *Frequency*: How often are cities (including mayors, city officials, councillors, community and business representatives and other urban interests) interacting with transnational city-networks?
- *Duration*: Over what periods of time do these interactions take place? How long have cities been involved in the networks?
- *Intensity*: What kinds of resources—in the form of time, money, etc.—are being transferred?
- *Institutional arrangements*: Finally, what are the terms of engaging with in transnational city-networks? What are the rules of obtaining and maintaining membership status? How do these rules affect frequency, duration and intensity of member contributions and engagements?

To summarize, there is now a growing body of empiricism that recognizes that transnational city-networks are playing a critical role in advocating and building capacity for climate governance at the urban scale. Empirical insights about the shift from advocacy and awareness raising to governance and implementation raise important theoretical questions about the factors affecting city engagement, the implications of city engagement on local capacity and the mechanisms through which city-networks—and influential actors within these networks—are able to project power at multiple scales of analysis (Acuto 2013; Gordon 2013; Gordon and Acuto 2015; Lee 2015; Hickmann 2016; Gordon and Johnson 2017).

At the same time, city-networks are clearly constrained by the resources and autonomy they are able to achieve in relation to corporations and the multilateral regime. As Thomas Hickmann (2016: 87) has observed:

> ICLEI neither pursues its own agenda separately from the UNFCCC process, nor does it aim to position transnational city-networks as alternative governance arrangements to multilateral treaty-making. By contrast, the analysis indicates that the ICLEI network directs its activities very much to the UNFCCC process.

Understanding the power of cities and city-networks therefore raises important questions about the ways in which corporations and private capital are shaping the politics of urban climate mitigation.

3.4 Corporate Power

In contrast to Westphalian and network power, corporate power implies the importance of material factors and relations (e.g. labour, resources and capital) in projecting influence through transnational networks and the UNFCCC. Power in this guise is not necessarily separate or distinct from Westphalian or network power, but it accentuates (and therefore draws our attention to) the role that multilateral donors and multinational corporations (MNCs) (e.g. Siemens, ARUP, the World Bank, the Green Climate Fund) play in shaping the actions and commitments of cities and city-networks on climate change.

As noted in Chap. 2, aid agencies and MNCs have come to play a critical role in developing and disseminating models of urban climate governance, including plans and metrics for reducing GHG emissions at the urban scale (Bouteligier 2012, 2015; Gordon and Acuto 2015; Arup-C40 2016). Among multilateral donors, the World Bank, the Asian Development Bank, the Inter-American Development Bank and the African Development Bank now offer a large number of urban climate mitigation programmes, many of which are working alongside transnational networks, such as ICLEI and the C40, and the UNFCCC. Similarly, multinationals like Siemens and ARUP have become increasingly involved in working with cities and city-networks (including the C40), designing climate-smart, low-carbon cities (Gordon and Acuto 2015; Bouteligier 2015; Arup-C40 2011, 2014, 2016).

Recognizing the role of multilaterals and MNCs in funding, designing, disseminating and implementing urban climate plans and policies raises a number of interesting questions about the changing nature of power in global climate politics. First, questions can be raised about the extent to which multilaterals and MNCs are shaping the content of urban climate policy actions and commitments. Second, and related, questions can be raised about the ways in which corporate models of urban climate governance are being used to measure and evaluate the performance of cities in relation to other financial governance metrics (cf. Gordon and Johnson 2017).

Arguably, one of the more important corporate actors now working on global climate policy is Arup (Box 3.1), a multinational consulting firm based in the UK. To date, Arup's principal involvement in urban climate policy has entailed a combination of political advocacy and technical assistance, involving substantial collaboration with the C40 on two editions of *Climate Action in Megacities* (Arup-C40 2011, 2014) and the more recent *Deadline 2020* (Arup-C40 2016). As Gordon and Acuto (2015: 69) have observed, ARUP and the C40 have collaborated extensively in "promoting and pooling" city capacity for reducing emissions while at the same time "tracking and measuring" the progress of individual city actions and commitment. By establishing emissions baselines and reporting protocols, ARUP and the C40 have been able to attract the investment of major international funders, such as Siemens, the World Bank Institute, Bloomberg Philanthropies and the Clinton Climate Initiative (Gordon and Acuto 2015).

Box 3.1 The "ARUP Effect"

ARUP was founded in 1946 by the Anglo-Danish engineer, Ove Arup (1895–1988), now specializing in a wide range of professional services, including engineering, urban planning, design, project management and management consulting. According to its Website, the company has 92 offices in 42 countries around the world. In 2016, its reported revenues were in excess of 1.2 billion pounds, accounting for 16 million pounds in profit (Arup 2016). Among some of its most notable projects are the Sydney Opera House, the Barbican Centre in London, the Millennium Bridge (also in London), Rajiv Gandhi International Airport in Hyderabad and the (infamous) Michael Lee-Chin Crystal at the Royal Ontario Museum in Toronto.

The implications of "orchestrating" the production and dissemination of urban climate data and knowledge we explore in due course (Sect. 3.4). Before doing so, we first look at the particular ways in which ARUP has organized and disseminated the production of knowledge for climate governance.

ARUP-C40's principal point of departure (2016) is the establishment of an emissions baseline that is used for estimating the total carbon budget that member cities have for staying within the 1.5-degree target. Drawing upon the Global Protocol for Community-Scale Greenhouse Gas Inventories (GPC, see below), *Deadline 2020* (Arup-C40 2016: 28) differentiates among "Scope 1" emissions, which include emissions that are produced within municipal boundaries, including "the direct emissions from combustion of fuels for heating, transportation etc.", Scope 2 emissions, which are the "indirect emissions" from power plants within the municipality, and a broader category of "Scope 3 emissions" that describe a city's indirect emissions that are associated with the consumption of food and materials that are produced outside of the municipality.

To stay within the 1.5-degree target, the authors estimate that average per capita emissions for the 84 cities that were members of the C40 in 2015 would have to be reduced "from over 5 tCO2e per capita today to around 2.9 tCO2e per capita by 2030" (IBID, 96). To put a financial price on it, the report then makes the case that a 2.1 tCO2e per capita reduction will require 14,000 additional "actions", which translates into roughly 140 actions per C40 city per year by 2020 (Arup-C40 2016: 42). Leaving aside the aforementioned challenge of associating actions with actual reductions, the importance of converting baselines into actions deemed necessary for maintaining carbon budgets is that it facilitates the calculation of costs for achieving the 1.5-degree target. According to the authors, the total cost of achieving an additional 14,000 actions by 2020 is US$375 billion, which is roughly equivalent to the (still unrealized) US$100 billion pledge to the Green Climate Fund.

An important part of Deadline 2020's emission reduction agenda is a proposed burden-sharing arrangement that differentiates cities on the basis of emissions projections and GDP per capita. Depending on where a city sits in relation to emissions and relative income, the idea is that cities will embark upon one of four mitigation trajectories:

- "Steep Decline"—For cities with a GDP per capita over $15,000 and emissions that are above the C40 average, emissions need to be immediately and rapidly reduced and the city is sufficiently developed to do so.
- "Steady Decline"—For cities with a GDP per capita over $15,000 but emissions that are lower than the C40 average, the city is sufficiently developed to immediately reduce emissions, but a less rapid rate of reduction is required than for the Steep Decline group.
- "Early Peak"—For cities with GDP per capita below $15,000 and higher than average emissions per capita, an early emissions peak is required, although the city's development status means that decline cannot be immediate.
- "Late Peak"—For cities with a GDP per capita below $15,000 and lower than average emissions per capita, a slightly later emissions peak is possible.

Leaving aside the (enormous) challenges of ensuring that member cities actually abide by the formula laid out in *Deadline 2020*, the proposed model of burden sharing raises significant questions about the practical meaning of terms like "a slightly later emissions peak" or "a less rapid rate of reduction". By setting the low threshold at $15,000 per capita, it also seems to exclude (or at the very least underplay) a large number of high-emitting cities (e.g. Dhaka, Delhi, Kolkata and Mumbai), whose GDP per capita falls well below the US$15,000 threshold.[13] Indeed, the member cities that are included (as examples) in the "low" category (Quito, Cape Town, Caracas, Durban and Amman) are ones whose relative affluence appears to confirm the assertion that the C40 is engaging primarily with wealthy and well-connected cities. Despite its laudable claim that "it is in all cities' interests to begin reducing per capita emissions as soon as possible" (Arup-C40 2016: 35), *Deadline 2020* provides little detail on how precisely it will encourage and achieve these ambitious policy targets and actions at the urban scale.

As a modelling exercise, questions can therefore be raised about the ways in which *Deadline 2020* is associating policy "actions" with actual

[13]https://www.brookings.edu/research/global-metro-monitor/. Last accessed 3 May 2017.

emission reductions. As a guide for future action, it identifies "five pathways" that may be used for achieving the 1.5-degree target:

- *The urban planning pathway*, involving land use policies that promote higher density living, containing urban sprawl;
- *The transit pathway*, emphasizing especially improvements in public bus services, "travel demand management", low-emission trucks and private vehicles and improvements in freight;
- *The energy pathway*, which includes scalable clean energy production, fuel and utility energy switching, district energy and improvements in energy efficiency;
- *The buildings pathway*, emphasizing new building codes and reporting protocols for new and existing buildings across all sectors; and
- *The waste pathway*, which involves recycling, education and awareness raising and full cost recovery for urban waste management.

Finally, the report identifies five ways of "unlocking" the potential of C40 member cities for meeting the 1.5-degree target:

- Engaging mayoral leadership, working with individual mayors to develop "a robust climate action plan" that is consistent with the aims of the Paris Agreement;
- Supporting cities to prepare robust climate inventories, targets and plans;
- Accelerating action through peer-to-peer exchange and ramping up direct support;
- Removing barriers to climate action; and
- Delivering global thought leadership, agenda setting communications and world-class events.

Apart from the fourth aim of "removing barriers to climate action", the proposed actions appear little different from those employed by other city-networks, such as ICLEI (cf. Hickmann 2016), and from earlier efforts on the part of the C40 to reduce emissions at the urban scale (cf. Gordon and Acuto 2015). What makes Number 4 interesting is that the barriers being specified are largely financial barriers, as opposed to the legislative and regulatory ones that so frequently prevent local authorities from acting effectively on climate change (cf. Gore 2015; Johnson et al. 2015). Unsurprisingly, the proposed solution to the barrier imposed by inadequate finances is also financial. "While the data is not yet complete", the report concludes,

> ...initial estimates suggest that the city-level actions necessary to deliver the Deadline 2020 vision across the C40 cities could require investment of over $1 trillion to 2050. Just under half of this is required by 2020. Average total investment across the C40 cities of over US$50 billion per year may be required up to and beyond 2030 to move onto a 1.5 degree trajectory. On a per-city basis, ... US$10-30 billion will be required by 2050 depending on the region, with African and South & West Asian cities needing the most. (Arup-C40 2016: 81)

To realize these figures, the report proposes using the C40 Cities Finance Facility (CFF) to provide $20 m by 2020 in the form of seed funding that will be used to "unlock and access to up to $1bn of additional capital funding, by providing the connections, advice and legal/financial support that enables cities to develop more financeable projects" (Arup-C40 2016: 78).

The implication here is that ARUP and the C40 are working towards the parallel goal of providing advocacy/support while at the same time creating the financial and institutional conditions for leveraging other sources of capital in the form of multilateral (e.g. Green Climate Fund and World Bank) and private-sector investment. Whether the goal of mobilizing US$50bn per year will ever be realized, the report is notable for its heavy emphasis on private-sector finance and investment and its relative lack of attention to the governance challenge of ensuring that other (i.e. private sector) investors adhere to the goal of investing in actions that have the ultimate end of reducing the urban climate footprint.

In this sense, *Deadline 2020* arguably has more to say about changing the planning, reporting and implementation practices of member cities than it does about redirecting contemporary patterns of trade and investment (a concession it appears to make when addressing the challenge of decarbonizing energy and electricity in Chapter 7 of the report). The implication then is that the inventories, targets and plans being put in place by the ARUP, the C40 and other transnational networks and MNCs constitute a form of (performative) power that institutionalizes and normalizes new forms of practice (Gordon and Johnson 2017).

3.5 Performative Power

Performative power implies that cities exert and subject themselves to power and influence by institutionalizing and normalizing themselves in relation to particular governance arrangements, such as those put in place by corporations, networks and international regimes. In the context of the UNFCCC, being recognized as a "Non-State Actor" or a "non-Party Stakeholder" confers legitimacy and status that cities (in general) and to particular Cities (such as Paris during COP21) can use to acquire resources, influence members and define policy agendas. Similarly, within the context of the C40, being designated an "innovator city" confers membership and access to resources, contacts and access to other networks, such as ARUP, the Green Climate Fund and the UNFCCC.

However, performative power implies more than simply achieving status and recognition within a particular governance arrangement. Rather, status and recognition are themselves repositories of power and meaning that entail following standards, metrics and methodologies that render cities observable and comparable. As Gordon and Johnson (2017: 13) have argued, power in this mode "rests in the use of mechanisms such as benchmarking and ranking schemes, which render complex and contested political domains legible and logical, and at the same time serve to suppress political contestation through a process of 'objectification' and normalization".

Box 3.2 describes the four-stage process that cities are required to follow in order to join the C40/Compact of Mayors. In one sense, the process entails demonstrating a willingness on the part of cities to adopt the goals and targets that are laid out in the Compact ("Stage 1: making a commitment"). In another, it entails adopting a methodology that renders cities observable and comparable under a particular set of norms, standards and expectations ("Stage 2: taking inventory", "Stage 3: creating measurable targets" and "Stage 4: Establishing an action plan").

Box 3.2: Reducing the Urban Carbon Footprint: The Global Covenant's Four-Stage Process[14]

Stage 1. Making a commitment A mayor may register on either of the Compact's partner reporting platforms—carbon*n* Climate Registry or CDP—or email a letter of intent to https://www.info@compactofmayors.org. As part of their commitment, cities agree to perform a series of key activities within three years from the date of their commitment. Following this submission, a city will be contacted by the Compact support team and will receive an official "Commitment" badge to display on local digital or print materials.

Stage 2. Taking inventory Within one year of its commitment, a city must assess the current impacts of climate change in his/her city. To do so, the city must: (1) Build and complete an overall greenhouse gas emissions inventory utilizing Compact standard; (2) Identify climate risks; and (3) Report on both via the CDP or carbon*n* questionnaires.

Stage 3. Creating measurable targets Within two years, the registered city must update its GHG inventory to include a breakdown of emissions by sources and sectors; set a target to reduce its GHG emissions; conduct a "climate change vulnerability assessment" using the Compact standard; and report in its chosen platform.

Stage 4. Establishing an Action Plan Within three years, a city's strategic action plan must show how it will deliver on its commitment to reduce greenhouse gas emissions and adapt to climate change.

Source https://www.compactofmayors.org/resources/

From an internal governance perspective, the four-stage methodology entails observing and evaluating individual city actions, providing a critical means of screening potential applicants, regulating compliance and (in theory) meting out sanctions when cities fail to comply with network standards and expectations. Externally, it provides a systematic model of recognition and certification that can then be used to pool and acquire other resources in the form of international finance and investment (Gordon and Johnson 2017). As Gordon and Johnson (2017: 705) have argued, the kind of "emergent orchestration" being displayed in this

[14]https://www.compactofmayors.org/resources/#tab-phase-1. Last accessed 31 May 2017.

instance "rests on the ability to establish those standards that become widely accepted amongst all actors located within a common domain".

Emergent orchestration also entails the act of voluntarily conforming to the standards and metrics that are used to evaluate and compare cities over space and time, and with one another. Take, for instance, *the Global Protocol for Community-Scale Greenhouse Gas Emission Inventories*, the GPC. Jointly developed by ICLEI, the C40 and the US-based World Resources Institute, the GPC outlines a systematic methodology for calculating and reporting GHG emissions by gas, scope, sector and subsector. Also employed in the aforementioned *Deadline 2020* report (Arup-C40 2016), the methodology entails differentiating among "Scope 1, 2 and 3" types of emissions that are measured within, as a consequence of and beyond the physical boundaries of a city. Beyond cities and communities, the GPC also offers standards and guidance methodologies for companies and other organizations (e.g. NGOs) interested in preparing a corporate-level GHG emissions inventory.[15]

By accounting for emissions that are embodied in urban consumption processes, Scope 2 and Scope 3 inventories provide an important way of estimating the total urban carbon footprint, as opposed to calculate the emissions that are only produced within a city's boundaries (Peters 2008; Peters et al. 2008). However, consumption-based accounting entails greater uncertainty in the sense that it entails establishing where a city's boundaries start and end, which embodied emissions can reasonably be tied to a city's consumption patterns and the more general challenge of estimating total emissions by gas, scope, sector and subsector (Erickson and Tempest 2015; Arup-C40 2016; Gordon and Johnson 2017). There is also the problem of double counting (ICLEI 2015: 19–20) when local bodies with overlapping boundaries (e.g. a city and a ward) both report the same emissions data (Erickson and Tempest 2015; Arup-C40 2016; Dahal and Niemelä 2017).

In spite of these challenges, important work is being done on the metrics and repositories of information being used to calculate the role of urban mitigation actions on reducing emissions (Singh and Kennedy

[15]http://www.ghgprotocol.org/corporate-standard. Last accessed 6 May 2017.

2015; Erickson and Tempest 2015; Arup-C40 2016; Dahal and Niemelä 2017).

Another intriguing example of emergent orchestration is *ISO 37120: 2014*, a set of city indicators that was developed in 2014 to measure service delivery and quality of life at the urban scale.[16] According to the ISO Website's description,

> ISO 37120 establishes a set of standardized indicators that provide a uniform approach to what is measured, and how that measurement is to be undertaken…

Included in the standard are 17 sets of indicators that report on economic performance (7 indicators), education (7), energy (7), environment (8), finance (4), fire and emergency response (4), governance (6), health (7), recreation (2), safety (5), shelter (3), solid waste (10), telecommunications and innovation (3), transportation (9), urban planning (4), wastewater (5) and water and sanitation (6). Particularly important from a climate governance performance perspective are indicators 7.4, which measures the percentage of urban energy consumption that derives from renewable sources, and 8.3 on greenhouse gas emissions per capita.

Among the benefits that can be obtained by purchasing an ISO certification, *ISO 37120: 2014* provides a standard on which individual cities can be measured and judged for internal and external evaluation and investment. In the words of the ISO,

> Standardized indicators enable cities to assess their performance and measure progress over time and also to draw comparative lessons from other cities locally and globally. They also help to guide policy, planning and management across multiple sectors and stakeholders.[17]

[16] https://www.iso.org/files/live/sites/isoorg/files/archive/pdf/en/37120_briefing_note.pdf. Last accessed 8 May 2017.

[17] https://www.iso.org/files/live/sites/isoorg/files/archive/pdf/en/37120_briefing_note.pdf. Last accessed 8 May 2017.

Alongside the World Bank's city creditworthiness initiative,[18] the UN-governed Cities Climate Finance Alliance,[19] the CDP (formerly the Carbon Disclosure Project)[20] and the aforementioned GPC, certification programmes like *ISO 37120: 2014* provide an important means of integrating and ordering cities in relation to corporations, networks and regimes, while at the same time subjecting them to processes of orchestration, observation and control (Gordon and Johnson 2017). The implication is that certification is itself a form of (performative) power that makes cities more attractive and observable to international markets and investors, including multilateral donors and MNCs (Gordon and Johnson 2017).

3.6 Concluding Remarks

This chapter has used four different models of (Westphalian, network, corporate and performative) power to understand the ways in which city engagement in global climate politics affects our understanding of power and politics in the international system. Its particular focus is on the ways in which city engagement in multilateral, network and corporate arrangements project new forms of power in the international system.

Explicit in the Paris Agreement is an expectation that multilateral donors (like the World Bank and Green Climate Fund), transnational city-networks (like the C40 and ICLEI) and multinational corporations (like Siemens) will be instrumental in providing (or funding) the systems, capital and expertise that are essential in building local capacity for NDC reporting and evaluation. An important function here is the role that donors, networks and MNCs play in pooling resources, sharing ideas and building capacity at the urban scale (Funfgeld 2015; Gordon and Acuto 2015; Lee 2015).

---

[18]http://www.worldbank.org/en/topic/urbandevelopment/brief/city-creditworthiness-initiative#Strategy. Last accessed 8 May 2017.

[19]http://www.citiesclimatefinance.org/wp-content/uploads/2015/11/CCFLA-Joining-Instructions.pdf. Last accessed 8 May 2017.

[20]https://www.cdp.net/en/cities-discloser. Last accessed 8 May 2017.

For many cities, transnational networks provide critical points of entry and access to finance, technology and expertise used in the development or low-carbon and climate-resilient strategies (Chu et al. 2016; Carmin et al. 2012; Gore 2015; Hughes and Romero-Lankao 2014; Funfgeld 2015; Romero-Lankao et al. 2015; Setzer et al. 2015). Engaging in networks—either through membership or through informal affiliations—clearly provides an important means through which cities may showcase new policy initiatives, attract and emulate best practices that may also further solidify the orchestration of transnational city-networks.

At the same time, engagement in transnational city-networks implies processes of orchestration that appear to be at odds with the voluntary, reciprocal and horizontal assumptions we typically use to classify and define network engagement. Insofar as cities now purchase or receive (through aid transfers) technical assistance from multinationals, transnational networks, donors and the UNFCCC, it appears likely that many of these policies will have the effect of replicating the practices of generic approaches to urban climate governance.

Is this necessarily a problem?

On the one hand, corporations, networks, donors and the UNFCCC are clearly filling a need that stems from a lack of domestic urban capacity and expertise (Archer et al. 2014; Bouteligier 2015; Chu et al. 2016; Carmin et al. 2012; Gore 2015; Hughes and Romero-Lankao 2014; Funfgeld 2015; Romero-Lankao et al. 2015; Setzer et al. 2015). On the other, they are perhaps working at odds with the processes of innovation and experimentation that are often ascribed to urban systems (Bulkeley et al. 2015). If cities are simply adopting corporate packages and policies, the scope for experimentation appears limited, although this of course is an empirical question.

Beyond the implications that standardized cookie-cutter approaches may have on local capacity (and democracy) are wider questions about the effects of urban integration—and orchestration—on the rigidity and vulnerability of large and interconnected urban systems. The following chapter considers these questions by exploring the politics of climate vulnerability and adaptation at the urban scale.

A final tension concerns the aggregating mechanisms that are reflected in the actions and effects of cities subjecting and integrating themselves into systems of surveillance and comparison. Explicit in the Paris Agreement is an expectation that multilateral donors (like the World Bank and Green Climate Fund), transnational city-networks (like the C40 and ICLEI) and multinational corporations (like Siemens) will be instrumental in providing (or funding) the systems, capital and expertise that are essential in building local capacity for NDC reporting and evaluation. An important function here is the role that donors, networks and MNCs play in pooling resources, sharing ideas and building capacity at the urban scale (Funfgeld 2015; Gordon and Acuto 2015; Lee 2015). For many cities, transnational networks have been shown to provide critical points of entry and access to finance, technology and expertise used in the development or low-carbon and climate-resilient strategies; (Chu et al. 2016; Carmin et al. 2012; Gore 2015; Hughes and Romero-Lankao 2014; Funfgeld 2015; Romero-Lankao et al. 2015; Setzer et al. 2015). Engaging in networks—either through membership or through informal affiliations—clearly provides an important means through which cities may showcase new policy initiatives, attract and emulate best practices that may also further solidify the orchestration of transnational city-networks.

References

Abbott, K.W., J.F. Green, and R.O. Keohane. 2016. Organizational ecology and institutional change in global governance. *International Organization* 70 (2): 247–277.

Acuto, Michele. 2013. *Global cities, governance and diplomacy: The urban link*. New York: Routledge.

Acuto, M., and S. Rayner. 2016. City networks: Breaking gridlocks or forging (new) lock-ins? *International Affairs* 92 (5): 1147–1166.

Anderton, K., and J. Setzer. 2017. Subnational climate entrepreneurship: Innovative climate action in California and São Paulo. *Regional Environmental Change*. doi:10.1007/s10113-017-1160-2.

Archer, Diane, Florencia Almansi, Michael DiGregorio, Debra Roberts, Divya Sharma, and Denia Syam. 2014. Moving towards inclusive urban adaptation: Approaches to integrating community-based adaptation to climate change at city and national scale. *Climate and Development* 6 (4), 345–356. doi:10.1080/17565529.2014.918868.

Arup-C40. 2011. *Climate Action in Megacities*. http://publications.arup.com/publications/c/climate_action_in_megacities. Last accessed 5 July 2017.

Arup-C40. 2014. *Climate Action in Megacities 2.0*. http://www.c40.org/blog_posts/CAM2. Last accessed 5 July 2017.

Arup-C40. 2015a. *Climate Action in Megacities 3.0*. http://www.cam3.c40.org/images/C40ClimateActionInMegacities3.pdf. Last accessed 14 June 2017.

Arup-C40. 2015b. *Potential for Climate Action*. http://publications.arup.com/publications/p/potential_for_climate_action. Last accessed 14 June 2017.

Arup-C40. 2016. *Deadline 2020: How cities will get the job done*. http://www.c40.org/researches/deadline-2020. Last accessed 14 June 2017.

Arup. 2016. *Financial Statements 2016*. http://publications.arup.com/publications/a/arup_financial_statements/arup_financial_statements_2016. Last accessed 5 July 2017.

Aykut, S.C. 2016. Taking a wider view on climate governance: Moving beyond the 'iceberg', the 'elephant', and the 'forest'. *Wiley Interdisciplinary Reviews: Climate Change* 7 (3): 318–328.

Aylett, Alex. 2015. Relational agency and the local governance of climate change: International trends and an American Exemplar. In *The urban climate challenge: Rethinking the role of cities in the global climate regime*, ed. Craig Johnson, Noah Toly, and Heike Schroeder, 156–180. New York and London: Routledge Press.

Bansard, J., O. Widerberg, and P. Pattberg. 2016. Cities to the rescue? Assessing the performance of transnational municipal networks in global climate governance. Paper presented at the 2016 International Studies Association, Atlanta, Georgia.

Barber, B. 2016. A governance alternative to faltering nation-states. *The Atlantic Citylab*. http://www.citylab.com/politics/2016/12/a-governance-alternative-to-faltering-nation-states/509612/?utm_source=nl__link4_120616. Last accessed 7 Dec 2016.

Barber, B. 2017. *Cool cities: Urban sovereignty and the fix for global warming*. New Haven: Yale University Press.

Barber, Benjamin R. 2013. *If mayors ruled the world: Dysfunctional nations, rising cities*. New Haven, CT: Yale University Press.

Bäckstrand, K., and J.W. Kuyper. 2017. The democratic legitimacy of orchestration: the UNFCCC, non-state actors, and transnational climate governance. *Environmental Politics* 26 (4): 764–788.

Bernstein, Stephen, and Benjamin Cashore. 2012. Complex global governance and domestic policies: Four pathways of influence. *International Affairs* 88 (3): 585–604.

Betsill, Michele, and Harriet Bulkeley. 2004. Transnational networks and global environmental governance: The cities for climate protection program. *International Studies Quarterly* 48 (2): 471–493.

Betsill, M., N.K. Dubash, M. Paterson, et al. 2015. Building productive links between the UNFCCC and the broader global climate governance landscape. *Global Environmental Politics* 15 (2): 1–10.

Biermann, F. 2011. Beyond the intergovernmental regime: recent trends in global carbon governance. *Current Opinion in Environmental Sustainability 2012* 4: 51–60.

Biermann, Frank, and Philipp Pattberg (eds.). 2012. *Global environmental governance reconsidered*. Cambridge: MIT Press.

Bouteligier, Sofie. 2012. *Cities, networks, and global environmental governance: Spaces of innovation, places of leadership*, Vol. 1. New York and London: Routledge.

Bouteligier, Sofie. 2015. Multinational companies and urban climate governance: Market making or successful policy innovation? In *The urban climate challenge: Rethinking the role of cities in the global climate regime*, ed. Craig Johnson, Noah Toly, and Heike Schroeder, 82–100. New York and London: Routledge.

Brenner, Neil. 2013. Theses on urbanization. *Public Culture* 25 (1): 85–114. doi:10.1215/08992363-1890477.

Brenner, Neil, and Christian Scmid. 2013. The 'urban age' in question. *International Journal of Urban and Regional Research*. doi:10.1111/1468-2427.12115.

Bulkeley, H. 2010. Cities and the governing of climate change. *Annual Review of Environment and Resources* 35: 229–253.

Bulkeley, H., and V. Castan Broto. 2012. A survey of urban climate change experiments in 100 cities. *Global Environmental Change* 23 (1): 92–102.

Bulkeley, H., and M. Betsill. 2013. Revisiting the urban politics of climate change. *Environmental Politics* 22 (1): 136–154.

Bulkeley, H. et al. 2014. *Transnational climate change governance*. Cambridge: Cambridge University Press.

Bulkeley, Harriet, Vanesa Castan Broto, and Gareth A.S. Edwards. 2015. *An urban politics of climate change*. London and New York: Routledge.

Burch, Sarah, Alison Shaw, Freya Kristensen, John Robinson, and Ann Dale. 2015. Urban climate governance through a sustainability lens: Exploring the integration of adaptation and mitigation in four british columbian cities. In *The urban climate challenge: Rethinking the role of cities in the global climate regime*, ed. Craig Johnson, Noah Toly, and Heike Schroeder, 119–138. New York and London: Routledge.

Carmin, J., I. Anguelovski, and D. Roberts. 2012. Urban climate adaptation in the global south: Planning in an emerging policy domain. *Journal of Planning Education and Research* 32 (1): 18–32.

Castan Broto, V. 2017. Urban governance and the politics of climate change. *World Development*. http://dx.doi.org/10.1016/j.worlddev.2016.12.031.

Chan, S., H. Asselt, T. Hale, K.W. Abbott, M. Beisheim, M. Hoffmann, and P. Pauw. 2015. Reinvigorating international climate policy: A comprehensive framework for effective nonstate action. *Global Policy* 6 (4): 466–473.

Chu, E., I. Anguelovski, and J. Carmin. 2016. Inclusive approaches to urban climate adaptation planning and implementation in the global south. *Climate Policy* 16 (3): 372–392.

Davis, D. 2016. Can mayors actually rule the world? *Citiscope*. http://citiscope.org/habitatIII/commentary/2016/12/can-mayors-actually-rule-world?utm_source=Citiscope&utm_campaign=a10e62b5f0-Mailchimp_H3_2016_12_05&utm_medium=email&utm_term=0_ce992dbfef-a10e62b5f0-118068273. Last accessed 7 December 2016.

de Flander, K. 2015. Closed cycles-open city. In *The urban climate challenge: Rethinking the role of cities in the global climate regime*, ed. C. Johnson, N. Toly, and H. Schroeder, 37–62. London and New York: Routledge.

Erickson, P., and K. Tempest. 2015. Keeping cities green: Avoiding carbon-lock-in due to urban development, SEI Working Paper No. 2015–11. *Stockholm Environment Institute: Seattle, WA, US*. https://www.sei-international.org/mediamanager/documents/Publications/Climate/SEI-WP-2015-11-C40-Cities-carbon-lock-in.pdf. Last accessed 14 June 2017.

Falkner, R. 2016. The Paris agreement and the new logic of international climate politics. *International Affairs* 92 (5): 1107–1125.

Finnemore, M., and K. Sikkink. 1998. International norm dynamics and political change. *International Organization* 52 (4): 887–917.

Funfgeld, H. 2015. Facilitating local climate change adaptation through transnational municipal networks. *Current Opinion in Sustainability* 12: 67–73. https://doi.org/10.1016/j.cosust.2014.10.011.

Geels, F. 2004. From sectoral systems of innovation to socio-technical systems: Insights about dynamics and change from sociology and institutional theory. *Research Policy* 33: 897–920.

Geels, F., and R. Raven. 2006. Non-linearity and expectations in niche-development trajectories: Ups and downs in Dutch biogas development (1973–2003). *Technology Analysis & Strategic Management*, 18 (3–4): 375–392. doi:10.1080/09537320600777143.

Geels, F. and J. Schot. 2007. Typology of sociotechnical transition pathways. *Research Policy* 36: 399–417.

Gordon, David J. 2013. Between local innovation and global impact: Cities, networks, and the governance of climate change. *Canadian Foreign Policy Journal* 19 (3), 288–307. doi:10.1080/11926422.2013.844186.

Gordon, David, and Michele Acuto. 2015. If cities are the solution, what are the problems? The promise and perils of urban climate leadership. In *The urban climate challenge: Rethinking the role of cities in the global climate regime*, ed. Craig Johnson, Noah Toly, and Heike Schroeder, 63–81. New York and London: Routledge.

Gordon, D.J. 2016. The politics of accountability in networked urban climate governance. *Global Environmental Politics* 16 (2): 82–100.

Gordon, D., and C. Johnson. 2017. The orchestration of global urban climate governance: Conducting power in the post-Paris climate regime. *Environmental Politics* 26 (4): 694–714.

Gore, Christopher. 2015. Climate change adaptation and African cities: Understanding the impact of government and governance on future action. In *The urban climate challenge: Rethinking the role of cities in the global climate regime*, ed. Craig Johnson, Noah Toly, and Heike Schroeder, 205–226. New York and London: Routledge.

Hale, T. 2016. "All Hands on Deck": The Paris agreement and nonstate climate action. *Global Environmental Politics* 16 (3): 12–22.

Hale, T., and C. Roger. 2014. Orchestration and transnational climate governance. *The Review of International Organizations* 9 (1): 59–82.

Hickmann, T. 2016. *Rethinking authority in global climate governance: How transnational climate initiatives relate to the international climate regime.* London and New York: Routledge.

Hochstetler, K., and M. Milkoreit. 2013. Emerging powers in the climate negotiations: shifting identity conceptions. *Political Research Quarterly* 67 (1): 224–235.

Hoffmann, Matthew J. 2011. *Climate governance at the crossroads experimenting with a global response after Kyoto*. Oxford: Oxford University Press.

Hsu, A., A.S. Moffat, A.J. Weinfurter, and J.D. Schwartz. 2015. Towards a new climate diplomacy. *Nature Climate Change* 5 (6): 501–503.

Hsu, A., Y. Cheng, A. Weinfurter, K. Xu, and C. Yick. 2016. Track climate pledges of cities and companies. *Nature* 532: 303–306. http://www.nature.com/news/track-climate-pledges-of-cities-and-companies-1.19764. 21 April 2016.

Hurrell, A., and S. Sengupta. 2012. Emerging powers, North–South relations and global climate politics. *International Affairs* 88 (3): 463–484.

Hughes, Sarah, and Patricia Romero-Lankao. 2014. Science and institution building in urban climate-change policymaking. *Environmental Politics* 23 (6), 1023–1042. http://dx.doi.org/10.1080/09644016.2014.921459.

Dahal, Karna, and Jari Niemala. 2017. Cities' greenhouse gas accounting methods: A study of helsinki, stockholm, and copenhagen. *Climate* 5 (31): 1–14.

ICLEI. 2015. *Carbonn climate registry: 5 year overview report: 2010–2015*. http://www.carbonn.org. Last downloaded 10 Sep 2016.

Johnson, Craig. 2017. Holding polluting countries to account for climate change: Is 'loss and damage' up to the task? *Review of Policy Research* 34 (1): 50–67.

Johnson, Craig, Noah Toly, and Heike Schroeder. 2015. Urban resilience, low carbon governance and the global climate regime. In *The urban climate challenge: Rethinking the role of cities in the global climate regime*, 3–23. New York and London: Routledge.

Kennedy, C. 2011. *The evolution of great world cities: Urban wealth and economic growth*. Toronto: University of Toronto Press.

Keohane, Robert, and David Victor. 2011. The regime complex for climate change. *Perspective on Politics* 9: 7–23.

Keohane, Robert, and David Victor. 2016. Cooperation and discord in global climate policy. *Nature Climate Change*. 9 May 2016. doi:10.1038/NCLIMATE2937.

Lee, Taedong. 2015. *Global cities and climate change: The translocal relations of environmental governance*. New York: Routledge.

Leichenko, Robin. 2011. Climate change and urban resilience. *Current Opinion in Environmental Sustainability* 3: 164–168.

McCann, E. 2011. Urban policy mobilities and global circuits of knowledge: Toward a research agenda. *Annals of the Association of American Geographers* 101 (1): 107–130.

Ostrom, E. 2010. Polycentric systems for coping with collective action and global environmental change. *Global Environmental Change* 20: 550–557.

Pahl-Wostl, C. 2009. A conceptual framework for analysing adaptive capacity and multi-level learning processes in resource governance regimes. *Global Environmental Change* 19: 354–365.

Paterson, M., M. Hoffman, M. Betsill, and S. Bernstein. 2014. Micro foundations of policy diffusion toward complex global governance: An analysis of the transnational carbon emission trading network. *Comparative Political Studies* 47 (3): 420–449.

Peters, G.P. 2008. From production-based to consumption-based national emission inventories. *Ecological Economics* 65 (1): 13–23. doi:10.1016/j.ecolecon.2007.10.014.

Peters, G.P., Edgar G. Hertwich. 2008. Post-Kyoto Greenhouse gas inventories: Production versus consumption. *Climatic Change*, 86: 51–66. doi:10.1007/s10584-007-9280-1.

Reams, M.A., K.W. Clinton, and N.S. Lam. 2012. Achievement of climate blanning objectives among US member cities of the international council for local environmental initiatives (ICLEI). *Low Carbon Economy* 3 (4): 137.

Romero-Lankao, P., and D. Dodman. 2011. Cities in transition: Transforming urban centers from hotbeds of GHG emissions and vulnerability to seedbeds of sustainability and resilience. *Current Opinion in Environmental Sustainability* 3: 113–120.

Romero-Lankao, Patricia, et al. 2015. Multilevel governance and institutional capacity for climate change responses in Latin American cities. In *The urban climate challenge: Rethinking the role of cities in the global climate regime*, ed. Craig Johnson, Noah Toly, and Heike Schroeder, 181–204. New York and London: Routledge.

Sassen, S., and N. Dotan. 2011. Delegating, not returning, to the biosphere: How to use the multi-scalar and ecological properties of cities. *Global Environmental Change* 21: 823–834.

Setzer, Joana, Laura Valente de Macedo, and Fernando Rei. 2015. Combining local and transnational action in the adoption and implementation of climate policies in the city of Sao Paulo. In *The urban climate challenge: Rethinking the role of cities in the global climate regime*, ed. Craig Johnson, Noah Toly, and Heike Schroeder, 101–118. New York and London: Routledge.

Singh, S., and C. Kennedy. 2015. Estimating future energy use and CO2 emissions of the world's cities. *Environmental Pollution* 203: 271–278.

Spector, J. 2015. The Paris climate deal can't succeed without cities. *The Atlantic Citylab.* http://www.citylab.com/cityfixer/2015/12/cop21-paris-climate-treaty-cities-success/420288/. Last accessed 7 Dec 2016.

Stevenson, Hayley, and John Dryzek. 2014. *Democratizing global climate governance*. Cambridge: CUP.

Tanner, T., T. Mitchell, E. Polack, and B. Guenther. 2009. Urban governance for adaptation: Assessing climate change resilience in ten Asian cities. IDS Working Paper 315, Institute of Development Studies, Brighton.

Tavares, R. 2016. *Paradiplomacy: Cities and states as global players*. Oxford: Oxford University Press.

Toly, Noah J. 2008. Transnational municipal networks in climate politics: From global governance to global politics. *Globalizations* 5 (3): 341–356.

van der Heijden, J. 2017. *Innovations in urban climate governance: Voluntary programs for low-Carbon buildings and cities*. Cambridge: Cambridge University Press.

van der Ven, H., S. Bernstein, and M. Hoffmann. 2017. Valuing the contributions of nonstate and subnational actors to climate governance. *Global Environmental Politics* 17 (1): 1–20.

Widerberg, O., and J. Stripple. 2016. The expanding field of cooperative initiatives for decarbonization: A review of five databases. *Wiley Interdisciplinary Reviews: Climate Change* 7: 486–500.

World Bank. 2010a. *Cities and climate change: An urgent agenda*. Washington, DC: World Bank. http://siteresources.worldbank.org/INTUWM/Resources/340232-1205330656272/CitiesandClimateChange.pdf. Retrieved 24 May 2017.

World Bank. 2010b. *The economics of adaptation to climate change*. Washington, DC: World Bank. https://siteresources.worldbank.org/EXTCC/Resources/EACC_FinalSynthesisReport0803_2010.pdf. Retrieved 24 May 2017.

Zenghelis, Dimitri, and Nicholas Stern. 2016. This is humankind's 'Great Urbanisation': We must do it right, or the planet will pay. *The Guardian*. November 8. https://www.theguardian.com/cities/2016/nov/08/mankind-great-urbanisation-era-act-now-planet-pay. Last downloaded 9 November 2016.

4

Resilient Cities? The Global Politics of Urban Climate Adaptation

Abstract This chapter explores the extent to which cities and city-networks have been able to articulate and negotiate new forms of power by demonstrating a need for recognition and assistance that may facilitate adaptation at the urban scale. Building upon the previous chapter, it makes the case that the politics of urban climate adaptation have occurred in reaction and response to a Westphalian system that effectively reifies the power of nation-states. However, animating, disseminating and in some cases contradicting the politics of multilateral climate governance is a parallel process of network and corporate governance, whose metrics, indicators and benchmarking exercises provide important forms of performative power that are being used to organize and orchestrate the ways in which cities, city-networks and corporations are defining resilience, vulnerability and risk in the Paris climate regime.

Keywords Urban climate adaptation · Vulnerability · Resilience International cooperation · Climate change insurance · The Green Climate Fund

C.A. Johnson, *The Power of Cities in Global Climate Politics*, Cities and the Global Politics of the Environment, https://doi.org/10.1057/978-1-137-59469-3_4

4.1 Introduction

Alongside the narrative that cities have been stepping into the void created by intransigent national governments is the narrative that cities are becoming increasingly vulnerable to climate change, and that this vulnerability constitutes a new form of power that is being used to articulate and legitimate recognition and assistance at the global scale. However, unlike the power that cities are projecting by setting, adopting and transcending international mitigation targets, power in this guise entails articulating, negotiating and demonstrating the need for the development of industries, assets, infrastructure and services that will facilitate adaptation to climate change.

By virtue of their size and geographic proximity to floodplains, river valleys and coastal zones, cities face multiple vulnerabilities that manifest themselves in the form of floods, heat waves, windstorms, landslides and the spread of water-borne disease (McGranahan et al. 2007; Munslow and O'Dempsey 2011; McDonald et al. 2011; Romero-Lankao and Dodman 2011; Mehrotra et al. 2011; Rosenzweig et al. 2011, 2015; Revi et al. 2014; Romero-Lankao et al. 2016). According to the 2011 Revision of the World Urbanization Prospects [UN/DESA (United Nations/Department of Economic and Social Affairs) 2012: 18], 39 of the 63 urban areas with populations greater than 5 million are located in areas with a high risk of flooding, cyclones or drought; 72% are located "on or near the coast", two-thirds are in Asia. According to another global assessment of 136 coastal cities (Hallegatte et al. 2013), the estimated annual cost of cities adapting to climate change will be in the range of US$350 million per city—or $US50 billion per year.

Building the resilience of cities to climate change is a global priority, but building urban resilience entails planning, regulating and investing in processes that frequently exceed the capacity of local governance bodies and institutions (Leichenko 2011; Atkins PLC 2012; Revi et al. 2014; Rosenzweig et al. 2011, 2015; Romero-Lankao et al. 2016; UNEP 2016; Arup-Rockefeller 2016). First, cities are often under great pressure to convert wetlands, river valleys, coastal areas and floodplains into residential, industrial, commercial and infrastructural real estate that, on the one hand, provides an important source of revenue (through business levies and property taxes) but, on the other, destroys the city's natural floodplains, reservoirs and carbon sinks [UNISDR (UN International Office for Disaster Risk Reduction) 2012; Revi et al. 2014; Romero-Lankao et al. 2016; Johnson et al. 2017]. Second, even the most well-endowed cities lack basic data on the location of flood zones, informal settlements, disease clusters and historical water levels that may be used in planning and regulating "safe" and sustainable urban practices [UNISDR (UN International Office for Disaster Risk Reduction) 2012; Satterthwaite and Bartlett 2017]. Third, cities are often subject to wider processes of economic globalization, land and property speculation, rural–urban migration, resource competition and pollution from other jurisdictions that undermine their ability to act effectively and autonomously on climate change (Leichenko 2011; Atkins PLC 2012; Revi et al. 2014; Rosenzweig et al. 2011, 2015; Romero-Lankao et al. 2016). Fourth, cities are themselves bound by past policy actions and decisions that lock them into particular modes of housing, planning, transportation and land use that exacerbate their vulnerability to climate change (Kennedy 2011; Leichenko 2011; Romero-Lankao et al. 2016). Finally, cities have in recent years been characterized by high levels of inequality, resulting from uneven processes of dispossession and displacement, that undermine the ability of marginal populations to respond, rebuild and adapt to rapid-onset disasters and long-term environmental change (Wisner et al. 2004; Satterthwaite et al. 2007, 2010; Wachsmuth et al. 2016; Castan Broto 2017; Florida 2017).

Taken together, such challenges raise important questions about the ability of cities to build resilience in a context of growing vulnerability to climate change. According to UNISDR (UN International Office for Disaster Risk Reduction) (2012: 11),

> A resilient city is characterized by its capacity to withstand or absorb the impact of a hazard through resistance or adaptation, which enable it to maintain certain basic functions and structures during a crisis, and bounce back or recover from an event.

Vulnerability, on the other hand, implies "a state of susceptibility to harm from exposure to stresses associated with environmental and social change and from the absence of capacity to adapt" (Adger 2006: 268).[1]

Within the urban context, definitions of urban vulnerability often emphasize the exposure of populations, private assets (e.g. housing and property) and public services and infrastructure to a variety of climate stressors (Atkins 2012; Romero-Lankao and Qin 2011; Romero-Lankao and Dodman 2011). The "Atkins Report" on "Future Proofing Cities" (Atkins 2012), for instance, conceptualizes vulnerability in relation to levels of poverty and inequality, the strength of basic services (in transportation, communication, sanitation, drinking water and healthcare) and urban form. Romero-Lankao and Qin (2011: 143–144) further distinguish between what they call "*vulnerability as impact*" studies, in which a primary focus is placed on "exposure to climate hazards, sensitivity of urban infrastructures, populations or activities, and the resulting or potential impacts" and "*inherent urban vulnerability*" approaches, which explore "how and why particular cities or populations are more vulnerable—or more able to cope or adapt than others".

A critical distinction therefore concerns the ways in which and extent to which vulnerability is conceptualized as "an end point of a linear process" or a "dynamic process based on the decreasing ability of a city or its populations to cope with a set of environmental hazards and stresses" (Romero-Lankao and Qin 2011: 145). Focusing on process highlights important questions about the historical-structural forces shaping urban vulnerability, including the historical policies (regulating land use, environmental protection, etc.) and patterns of trade and investment that induce large numbers of people to live and work in areas that are chronically (and in the case of climate change increasingly) exposed to climatic hazards (Adger 2006; Wisner et al. 2004; Satterthwaite et al. 2007, 2010; Castan Broto 2017).

---

[1]Exposure in turn implies the existence of social and biophysical pathways that connect climatic/environmental stressors with affected populations (Smit and Wandel 2006; Adger 2006). Finally, adaptive capacity implies the assets, public services, infrastructure and institutions (including formal services and informal coping mechanisms) that affected populations have at their disposal.

Focusing on process also highlights the ways in which past struggles, decisions, changes and "ruptures" affect contemporary development pathways.

From a resilience perspective, the development of roads, housing, infrastructure and public services all create potential rigidities that have strong bearing on the ability of urban systems to act and react to changing environmental conditions (cf. Sassen and Dotan 2011; Kennedy 2011; Sovacool 2011; Romero-Lankao et al. 2016). Whether cities can build resilience and reduce vulnerability at multiple scales depends on many factors, including the ability of municipalities, businesses, local communities and civil society organizations to invest in infrastructure and institutions that can be replicated and maintained in the face of future social and environmental stress [World Bank 2010a; Leichenko 2011; Melhotra et al. 2011; Atkins 2012; UNISDR (UN International Office for Disaster Risk Reduction) 2012; Arup-Rockefeller 2016; Romero-Lankao et al. 2016]. According to UNISDR (UN International Office for Disaster Risk Reduction) (2012: 11),

> Sound development practice with good regulations, well-maintained infrastructure, capable emergency management and solid institutions, which develop participatory urban plans, provide building permits, and manage water resources and solid waste, help to build up cities' resilience over time.

However, resilience as a concept has been criticized for overstating the value of recovering and "bouncing back", as opposed to addressing and understanding the structural factors that perpetuate poverty and vulnerability to environmental change (Romero-Lankao et al. 2016; Castan Broto 2017). In theory, "sound development practice", "good regulations" and "capable emergency management" are important principles that may be used in planning and restricting the construction of roads and buildings in high-risk areas (e.g. low-lying coastal zones, floodplains), improving the quality and availability of housing, clean water and medical services, and communicating disaster risk [Wisner et al. 2004; UNISDR (UN International Office for Disaster Risk Reduction) 2012]. In practice, the ability of cities to undertake these actions varies enormously, reflecting the aforementioned patterns of urbanization and socio-economic disparities, both within and among cities (Leichenko 2011; Atkins PLC 2012; Revi et al. 2014; Rosenzweig et al. 2011, 2015; UNISDR (UN International Office for Disaster Risk Reduction) 2012; Romero-Lankao et al. 2016; Arup-Rockefeller 2016; Castan Broto 2017). Overcoming the rigidities put in place by years of investment (or lack thereof) in particular modes of transportation, sanitation, land use planning and water governance entails an ability to assume the costs and potential risks of maintaining the status quo and an ability to mobilize (public and private) resources in the name of (long-term) infrastructural development (cf. Moser and Luers 2008; World Bank 2010a; Kennedy 2011; Revi et al. 2014).

Climate policy decisions also frequently entail difficult policy trade-offs between adaptation and mitigation (Burton [2004 (2009)]; Revi et al. 2014; Castan Broto 2017). Although policies aimed at reducing or phasing out large GHG emitters like coal-fired power plants or diesel-powered transportation can provide important co-benefits in the form of cleaner air and improved respiratory health, the perceived benefits of mitigating GHGs are often far removed from what are perceived to be the more pressing and immediate needs that often fall under the heading of urban adaptation. In the words of the World Bank:

> Adaptation will have impacts primarily on a local scale: actions are based on specific needs of the affected regions. Costs might be very high, especially in large-scale infrastructure such as flood protection works, roads, ports, and power generation facilities… but savings from adaptation efforts mainly accrue through reduced risk impacts (for example, insurance) often at an individual homeowner, business, or community level. (World Bank 2010a: 11)

Mitigation, by contrast:

> . . . is a global effort requiring broad changes of behavior and technological advancements. Mitigation strategies are usually expensive in the short term, because they are capital intensive (change in technology, urban transport, and collective infrastructure) and require fundamental changes to urban systems. Over time, the cost of mitigation is generally self-financed through cost savings (mainly in energy bills).

Here it is worth noting too that the information that city leaders, public officials, businesses and local communities use to support mitigation—as opposed to adaptation—is more dependent upon a scientific discourse that is mediated by experts, epistemic communities and "boundary organizations", such as the IPCC, whose observations and long-term projections play a critical role in shaping local mitigation plans and agendas (e.g. the 1.5-degree target of the Paris Climate Agreement). Although urban planning for adaptation, disaster risk reduction and resilience is also influenced by scientific and technical bodies (see below), the critical point here is that mitigation policy is more heavily dependent upon these highly specialized bodies of knowledge, encompassing climate modelling, atmospheric systems, feedback mechanisms and long-range projections (e.g. Grundmann 2013; Hoppe et al. 2013).

That being said, cities and non-state actors have thus far been far more likely to engage in mitigation than adaptation. According to ICLEI (2015: 21), 4709 (or 76%) of all of the 6181 climate policy actions that have been reported to the registry as of 2015 were classified as mitigation; 1472 were classified as adaptation.

Empirical insights about the economic and political challenges of building urban climate resilience have generated a number of important policy discussions about the mechanisms by which cities and city-networks may be able to overcome the "adaptation deficit" that exists between current local capacity and future vulnerability to climate change (World Bank 2010a; Rosenzweig et al. 2011; Atkins 2012; Revi et al. 2014; Arup-Rockefeller 2016; UNEP 2016). Within the multilateral regime, considerable attention has been paid to the prospects of using new sources of international climate finance (e.g. in the form of the Green Climate Fund) to fund the construction of sea walls, drainage systems and other forms of urban infrastructure that might be used in reducing the vulnerability of urban populations to climate change (Revi et al. 2014). Among city-networks, considerable attention has been paid to the challenge of incorporating marginal populations (e.g. homeless people, slum dwellers, migrant workers) into decision-making processes, in theory providing a more effective means of capturing the risks and vulnerabilities that are often under-represented in government and corporate plans and policies (Archer et al. 2014; Carmin et al. 2012;

Chu et al. 2015; Haque et al. 2014; Roy et al. 2013). Finally, and arguably contradicting this last sentiment, there is also a growing recognition that local, multilateral and network approaches to governing urban climate vulnerability are going to require extensive involvement and investment on the part of the private sector (Arup-Rockefeller 2016; UNEP 2016).

Arguably, the dominant narrative that has been used to frame multilateral action on urban climate governance is the notion that cities can build resilience by supporting "adaptation" at the urban scale. The "Urban Areas" chapter (Revi et al. 2014) of the IPCC's Fifth Assessment Report, for instance, highlights "the importance of city and municipal governments acting now to incorporate climate change adaptation into their development plans and policies and infrastructure investments".

> This includes not only building that foundation of resilience (and its institutional, governance and financial underpinnings) but also mobilizing new resources, adjusting building and land-use regulations and continuously developing the local capacity to respond.[2]

Similarly, Article 7.1 of the Paris Climate Agreement states that:

> Parties hereby establish the global goal on adaptation of enhancing adaptive capacity, strengthening resilience, and reducing vulnerability to climate change, with a view to contributing to sustainable development and ensuring an adequate adaptation response in the context of the temperature goal referred to in Article 2.

An important part of the Agreement was a formal commitment on the part of developed countries to provide US$100 billion per year in additional climate finance for funding adaptation (and mitigation) efforts in the Global South. As we shall see, the $US100 billion figure has become an important political symbol of "common but differentiated

[2]http://ipcc-wg2.gov/AR5/images/uploads/WGIIAR5-Chap8_FGDall.pdf. Last accessed 10 May 2017.

responsibility" within the multilateral climate regime. But it also represents an important article of faith that has been used to bring cities, city-networks, corporations and nation-states into the adaptation regime complex.

According to the IPCC, adaptation entails a wide range of "initiatives and measures to reduce the vulnerability of natural and human systems against actual or expected climate change effects".[3] Adaptation can take the form of *incremental* adjustments in which cities "maintain the essence and integrity of a system or process at a given scale", or *transformational adaptation*, which entails changing "the fundamental attributes of a system in response to climate and its effects".[4]

Distinguishing between incremental and transformational adaptation helps to differentiate the adaptive strategies that replicate existing policies, institutions and practices (i.e. "bouncing back"), as opposed to ones that challenge and change the status quo. Brooks et al. (2009: 753), for instance, argue in favour of challenging the status quo on the grounds that contemporary adaptation pathways "run the risk of locking societies into unsustainable and maladaptive patterns of development". Similarly, and more squarely rooted in the urban context, Shi (2016) makes the case that

> ... most adaptation proposals try to protect existing development in coastal and low-lying urban areas in ways that perpetuate continued growth in these exposed areas. It remains unclear how large proposed infrastructure projects such as the Great Garuda Seawall in Jakarta, Indonesia, or Eko Atlantis Island in Lagos, Nigeria, will affect vulnerable groups. The fact is, there are winners and losers in urban climate adaptation projects, and it is the poorest and most marginalized who (as always) tend to lose.

---

[3]Downloaded 17 December 2012 at http://www.ipcc.ch/publications_and_data/ar4/syr/en/annexessglossary-a-d.html.

[4]http://www.ipcc.ch/pdf/assessment-report/ar5/wg2/WGIIAR5-AnnexII_FINAL.pdf. Last accessed 10 May 2017.

Shi's comments underscore the possibility that local, national and international efforts to build urban resilience may actually or inadvertently foster *maladaptive* development pathways that make marginal and vulnerable urban populations more vulnerable to climate change. To take but one example, public and private insurance policies often create strong incentives to rebuild lives, homes and livelihoods in areas (e.g. hillsides and flood plains) that are chronically vulnerable to extreme weather events (Wisner et al. 2004; Sovacool et al. 2015). In the words of Canada's Globe and Mail:

> The impact of changing weather patterns is being compounded by antiquated municipal sewer systems that can't handle giant surges in water flow, and by the mind-blowing fact that homes are still being built in known flood plains.[5]

On this basis, further conceptual distinctions can therefore be made between "successful" adaptation, which implies an ability to learn from past experiences and to coordinate actions in ways that challenge or change existing behaviours (Holling 2001; Folke et al. 2005; Folke 2006; Olsson et al. 2006; Smit and Wandel 2006; Armitage et al. 2008; Pahl-Wostl 2009), and "maladaptation", which implies that actors (e.g. cities, city leaders, urban populations) are unable to change their behaviour to manage changing circumstances and events (Pahl-Wostl 2009: 358). In Pahl-Wostl's words, it implies an "inability . . . to develop expectations, coordinate collective action and improve routines and practices" (Pahl-Wostl 2009: 358).

Differentiating between adaptive and maladaptive development pathways helps to specify the normative terms on which we might define urban adaptation, resilience, vulnerability and risk (Wisner et al. 2004; Sovacool 2011; Ribot 2014; Sovacool et al. 2015). Sovacool (2011: 1177), for instance, distinguishes between "hard" and "soft" adaptation

[5]http://www.theglobeandmail.com/opinion/editorials/the-floodwaters-are-rising-in-canada-when-will-ottawa-finally-deal-with-it/article34969402/?reqid=fe5d6b86-4aaa-4968-90a2-df8aaeb7a458. Last accessed 13 May 2017.

pathways, the former encompassing "capital-intensive, large, complex, inflexible technology and infrastructure", and the latter prioritizing "natural capital, community control, simplicity and appropriateness". Within the urban context, examples of hard adaptation include climate proofing transportation networks and electrical grids, upgrading the building stock that is used in the construction of homes and businesses, and strengthening regulatory codes and standards for housing, transportation and sanitation (Rosenzweig and Solecki 2014). "Soft" adaptation, on the other hand, entails modifying (as opposed to radically changing) norms and behaviour in ways that reduce vulnerability, for example by pooling resources for community-based insurance (Sovacool 2011; UNEP 2016).

Critical reflections on the normative/policy implications of pursuing adaptive versus maladaptive development pathways highlight the prospects of fostering or engaging in more autonomous forms of "social resilience", whereby local communities and vulnerable populations are able to develop their own ways of building resilience to climate change (Sovacool 2011; Sovacool et al. 2015; Chu et al. 2017). As noted in Chap. 2, recent empirical work on the politics of urban climate governance suggests that cities can provide important sites of experimentation, in which new plans, codes and policies are used to reduce emissions and vulnerability at the urban scale (Hoffman 2011; Bulkeley and Castan Broto 2012; Anguelovski et al. 2014; Bulkeley et al. 2015). Framed in this way, cities can be usefully conceptualized as "complex adaptive systems", in which the convergence of social, public and private power creates new forms of innovation and resilience at the urban scale (Hoffman 2011; Bulkeley and Castan Broto 2012; Anguelovski et al. 2014; Bulkeley et al. 2015). Bulkeley and Castan Broto (2012: 1), for instance, find that climate policy experiments "serve to create new forms of political space within the city, as public and private authority blur, and are primarily enacted through forms of technical intervention in infrastructure networks".

However, much remains to be known about the theoretical, empirical and normative dimensions of urban climate policy experiments. First, notwithstanding the isolated instances in which cities have been able to create the conditions for climate-resilient development pathways, the vast majority of writing about urban climate policy experiments has focused on larger and wealthier "global cities", whose access to public finances, private investment and international climate policy networks makes them more visible in the extant literature on cities and climate change (Acuto 2013; Gordon 2013; Gordon and Acuto 2015; Carmin et al. 2012; Chu et al. 2015; Gordon 2013; Gore 2015; Johnson et al. 2015; Lee 2015; Leichenko 2011). Second, theoretical and empirical insights about the potential "ripple" effect of isolated policy experiments on wider, systemic patterns of resource consumption and governance remain thin (Bulkeley et al. 2015). Finally, questions have been raised about the extent to which these autonomous or semi-autonomous models of social resilience are simply delegating or deferring new forms of responsibility to local communities and vulnerable populations (Satterthwaite and Bartlett 2017).

In what follows I make the case that cities, city-networks and multinational corporations operating within the context of the multilateral regime have entered a new phase of interaction and interdependence in the global politics of climate adaptation. Much as we have seen in the case of urban climate mitigation (Chap. 3), the politics of urban climate adaptation have occurred in reaction and response to the Westphalian system of nation-states. However, animating, disseminating and in some cases contradicting the politics of multilateral climate governance is a parallel process of network governance and MNCs, whose metrics, indicators and benchmarking exercises provide important forms of performative power that are being used to organize and orchestrate the ways in which cities, city-networks and corporations are defining resilience, vulnerability and risk in the Paris climate regime.

4.2 Westphalian Power: Urban Resilience and the Multilateral Regime

By far, one of the most contentious aspects of contemporary climate politics has been the politics of financing developing (or in the language of Kyoto, "non-Annex I") countries for the purposes of adaptation and mitigation (Hurrell and Sengupta 2012; Hochstetler and Milkoreit

2013; Caravani et al. 2014; Falkner 2016; UNEP 2016). Although adaptation and mitigation have by and large been treated as separate research and policy categories within the framework convention, they have come together in relatively dramatic ways during the COPs. Prior to COP21 in 2015, the history of climate negotiations had been one in which industrialized countries (i.e. the USA) were generally reluctant to commit themselves to ambitious emissions reductions unless or until developing countries (i.e. China and India) agreed to do the same. Developing countries on the other hand were generally reluctant to commit themselves to binding emissions reductions unless they received a credible assurance that wealthier, industrialized countries would provide the necessary time and resources that they would need to de-couple from carbon-based models of economic development (Hurrell and Sengupta 2012; Hochstetler and Milkoreit 2013; Falkner 2016).

Underlying these and other climate finance initiatives is an assumption that more affluent countries have a responsibility to rectify the past and potential losses occurring as a result of climate change, ideally through the transfer of financial resources. According to the UNFCCC (Art. 2, cited in Schalatek and Bird 2014: 1), parties to the convention are expected to take actions, including on finance, on "the basis of equity and in accordance with their common but differentiated responsibilities and respective capabilities". Moreover, Article 4.3 of the convention states that climate financing should uphold principles of "adequacy and predictability in the flow of funds and the importance of appropriate burden sharing among the developed country parties" (cited in Schalatek and Bird 2014: 1).

Prior to Paris, it became readily apparent that developing countries were using future emission reductions commitments as a means of negotiating stronger commitments in the form of "new and additional" resources and finance. During the COP19 meetings in Warsaw, for instance, China and the G77 (and a number of high profile NGOs) led 132 countries in walking out of the talks on loss and damage after it became clear that the $100 billion that had been pledged in support of the Green Climate Fund was not materializing. Following Paris, developing countries appear to have achieved their goal of securing new climate financing commitments and a (relatively empty) commitment on the part of COP21 to "avert, minimize and address" the losses and damages associated with the adverse effects of climate change (Box 4.1).

Box 4.1 Compensation beyond adaptation? The politics of "loss and damage"

Formally established at COP19 in 2013, Loss and Damage represents what is for many an important effort on the part of developing countries (including China and the G77) to address the past and potential harms incurred as a result of climate change (Warner et al. 2012, 2013; Roberts et al. 2014; Zommers et al. 2014; Serdeczny et al. 2016; Johnson 2017). Underlying the mechanism is an assumption that identifying past and potential harms will provide a means of assigning responsibility for the inequitable nature of climate change, whose impacts will be most directly felt in the Global South (Shue 2010; Warner et al. 2012, 2013; Roberts et al. 2014; Zommers et al. 2014; Serdeczny et al. 2016; Johnson 2017). Among the principal activities currently being supported under the Loss and Damage mechanism are efforts to (1) enhance the knowledge and understanding of comprehensive risk management approaches to address loss and damage associated with the adverse effects of climate change, including slow onset impacts; (2) strengthen the dialogue, coordination, coherence and synergies among relevant stakeholders; and (3) enhance action and support, including finance, technology and capacity-building, to address loss and damage associated with the adverse effects of climate change (UNFCCC 2015a). By making specific reference to loss and damage in Article 8, the Paris Climate Agreement marks an important milestone for the Loss and Damage mechanism, thereby "ensuring the permanence of the issue under the UNFCCC" (Serdeczny et al. 2016: 6). However, its ability to uphold this principle has been constrained by the weakness of the UNFCCC and the obduracy of powerful nation-states (cf. Victor and Keohane 2011, 2016; Stevenson and Dryzek 2014). Article 52 of the Paris Agreement, for instance, states that party recognition of the Loss and Damage mechanism "does not involve or provide a basis for any liability or compensation", (UNFCCC 2015b: Article 52).

The UNFCCC currently provides a number of channels for financing climate change adaptation, including the Least Developed Country Fund (LDCF), which is intended to help with the identification of adaptation priorities in the National Adaptation Programmes of Action (NAPAs); the Special Climate Challenge Fund (SCCF), which is designed to support adaptation, as well as mitigation and technology transfer; the Adaptation Fund, which is aimed at supporting "concrete adaptation" actions and, most recently, the Green Climate Fund, which was established in 2010 to provide $US100 billion in additional climate financing by the year 2020.[6] All of these were developed in advance of the Paris Agreement, but it is the $100 billion that has come to pre-occupy contemporary discussions about the politics of adaptation (Shi 2016; UNEP 2016).

Formally established in 2010, the Green Climate Fund (GCF) is responsible for allocating resources "to low-emission and climate-resilient projects and programmes in developing countries". According to its Website, "the Fund pays particular attention to the needs of societies that are highly vulnerable to the effects of climate change, in particular Least Developed Countries (LDCs), Small Island Developing States (SIDS), and African States".[7] By the end of 2015, the GCF had approved the disbursement of US$35 billion for climate change adaptation, representing 76% of the resources that have been pledged for adaptation (UNEP 2016: xiii).

For many observers, the GCF represents a critical source of financial and technical support for urban adaptation (and mitigation) in the Global South. In practice, it faces a number of serious hurdles. First, the money that has been pledged in support of climate change adaptation has thus far been well below what is needed (Caravani et al. 2014; Shi 2016; UNEP 2016). According to the *Adaptation Finance Gap Report* (UNEP 2016: xii), a document that was prepared by the United Nations Environment Programme in advance of COP21, the total cost of adapting to climate change is in the range of US$140 billion to US$300 billion per year by 2030, rising to US$280 billion to US$500 billion by

[6]Outside of the UNFCCC, many donors and aid agencies are now actively involved in promoting adaptation programs of their own. Particularly notable from a multilateral perspective are *the City Resilience Profiling Programme* (https://unhabitat.org/urban-initiatives/initiatives-programmes/city-resilience-profiling-programme/) and the *Cities in Climate Change Programme* (https://unhabitat.org/urban-initiatives/initiatives-programmes/cities-and-climate-change-initiative/), both administered by UN-Habitat. For a comprehensive summary of international adaptation (and mitigation) initiatives, see "Climate Funds Update" at http://www.climatefundsupdate.org/data/the-funds-v2.

[7]http://www.greenclimate.fund/about-gcf/global-context#mission. Last accessed 16 May 2017.

2050 (Box 4.2). By contrast, the total amount of bilateral and multilateral funding that was allocated for climate change adaptation in 2014 was US$25 billion, US$22.5 billion of which was targeted at developing countries (UNEP 2016: xiii).

Although the Paris Agreement now reinforces the commitment on the part of industrialized countries to provide $US100bn per year by 2020, questions remain about whether and in what form the money will materialize. In the words of UNEP (2016: 22),

> …there is no agreement as to the type of funding that shall be mobilised to meet this goal, and financing is expected to come from "…a wide variety of sources, public and private, bilateral and multilateral, including alternative sources" ...This uncertainty has hampered efforts to monitor progress toward meeting the goal, despite recent efforts to improve tracking for climate finance.

Box 4.2 Adapting to climate change: How much is it going to cost?

Underlying the idea that wealthier countries have a responsibility to pay for climate change adaptation is an assumption that wealthier countries can identify and potentially rectify the costs of adapting to climate change. However, putting a price on the relative costs of action and inaction is by no means an exact science (World Bank 2010b; UNEP 2016). UNEP's estimate (2016), for instance, puts the cost at between US$140 billion and US$300 billion per year by 2030. The World Bank's (2010b) report on *the Economics of Adaptation to Climate Change*, on the other hand, put the annual cost of climate change adaptation at between US$70 and US$100bn per year between 2010 and 2050 in 2005 prices.

One of the challenges of estimating the cost of adaptation (and mitigation, for that matter) involves the assumptions we use in determining the value of future costs and benefits, including "trade-offs between the impacts of climate change, the costs of adaptation, and the residual costs after adaptation" (UNEP 2016: 15). On this point, the Stern Review was criticized (e.g. by Nordhaus 2007, among others) for overstating the benefits (or in the language of economics, understating the discount rate we use in calculating the value of future consumption) of future abatement efforts. Another challenge entails deciding what to include in the relevant range of indicators for adaptation. UNEP (2016: 5), for instance, includes private and autonomous adaptation, which it defines as "actions by households, businesses or communities acting on their own without public intervention, but within an existing public policy framework". Clearly, expanding the definition has implications for the value of the estimate. As UNEP (2016: 14) points out, "Studies with greater coverage will produce higher estimates, as they include a larger number of impacts". A related challenge entails aggregating disparate costs and values into regional and global estimates. Many adaptation estimates are based on formulae that calculate the larger cost (or benefit) of a particular action at regional and global scales of analysis. However, such calculations are highly dependent upon the validity and reliability of available data. Although the number of studies is growing, widening the potential database, estimates are still constrained by the quality and availability of reliable data. UNEP (2016: 2), for instance, distinguishes between "bottom-up" and "top-down" studies. "Bottom-up studies", it suggests,

> calculate costs by adding up the costs of each of the measures in a specific, pre-determined portfolio of adaptation actions. Typically, these actions are national or sub-national in scope. In contrast, top-down studies calculate costs by relating total impacts with impact damages, often at the global level and on the basis of a sectoral breakdown of cost elements.

A final and related challenge concerns the assumptions we make about future emissions scenarios, development pathways and economic growth projections, all of which are inherently uncertain.

Second, most of the adaptation funding that has been allocated to date has gone to rural areas and agricultural sectors, as opposed to cities and urban sectors (such as transportation, housing and urban sanitation). According to Revi et al. (2014: 588), a mere 20% of all climate adaptation funding had gone to urban areas in 2010–2011, which is well below the total number of people living and working in urban areas (Birkmann et al. 2010; Revi et al. 2014; Johnson et al. unpublished). In their study of 19 coastal Asian countries, Johnson et al. (unpublished) found that less than 10% of all national adaptation policy actions and commitments to the UNFCCC were directed at cities and urban issues. Although it contains an Urban Areas chapter in its Fifth Assessment Report (Revi et al. 2014), the IPCC makes no mention of "cities", "municipalities" or "urban" in its most recent (ca. 2017) glossary of terms.[8]

[8]http://www.ipcc.ch/pdf/assessment-report/ar5/wg2/WGIIAR5-AnnexII_FINAL.pdf. Last accessed 10 May 2017.

Explanations for the apparent "rural bias" in multilateral climate policy point to a number of factors, including the prominence of agricultural and rural development agencies in national climate policy (Birkmann et al. 2010; Revi et al. 2014), competing climate mitigation agendas (Revi et al. 2014) and the impact of rural domestic interest groups, such as farmers, agricultural labourers and the rural poor, on national climate policy processes and agendas (Birkmann et al. 2010; Revi et al. 2014).[9] Observations have also been made that cities have found it much easier to promote mitigation over adaptation agendas (Revi et al. 2014; Johnson et al. 2015), reflecting the larger "sunken" costs of investing in climate-resilient infrastructure programmes (e.g. climate proofing entire transportation networks as opposed to implementing a single landfill methane capture facility) and the heavy emphasis that has been placed on mitigation by the UNFCCC (Revi et al. 2014; UNEP 2016). As Revi et al. have observed (2014: 564):

> Local decision makers frequently view climate change as a marginal issue, but adaptation usually ranks lower than mitigation on the agenda.

In the light of these observations, increasing attention has been paid to the challenge of raising the political and epistemic profile of cities and urban populations in the multilateral regime (Rosenzweig et al. 2011; Anguelovski et al. 2014; Revi et al. 2014; Shi 2016; Shi et al. 2016; Chu et al. 2017). Driving this process, cities and city-networks (such as ICLEI, the C40 and the Urban Climate Change Research Network—UCCRN) have used a variety of programmes, conferences (e.g. ICLEI's annual *Resilient Cities* conference), publications and social media campaigns to raise awareness about the particular needs and vulnerabilities

[9] Note that these interests may also be ascribed to these groups, as when - for instance - national policy elites associate rural poverty with climate change adaptation.

of cities to climate change (see below). Increasing scholarly attention has also been paid to documenting the politics of urban adaptation policy processes in the Global South (e.g. Anguelovski et al. 2014; Revi et al. 2014; Shi 2016; Shi et al. 2016; Chu et al. 2017).

The result has been a groundswell of academic and popular writing about the prospects (and perils) of building climate resilience at the urban scale. However, many researchers, networks and city leaders are coming to the conclusion that the prospects for raising the profile of cities in the multilateral framework are to a very large degree constrained by the economic and political disparities that underlie the Westphalian system of nation-states. According to Shi (2016),

> The Paris talks accentuated the disparities between the capacities of different cities and peoples to adapt. Developed countries pledged $248 million in additional adaptation funding to be disbursed among the world's 48 least developed nations, home to 932 million people.

Much like aid giving more generally (cf. Riddell 2008), a major concern is that the lines of accountability and responsibility between donors and recipients of adaptation finance are blurred by a lack of clear criteria about what kinds of entitlements cities and urban populations are expected to receive and the conditions under which they may expect to receive them. Multilateral climate policies and programmes have long been criticized for being overly top-down and technocratic, promoting plans that reflect the priorities of donors, national governments and the UNFCCC, as opposed to vulnerable populations themselves (Forsyth 2013; Ayers et al. 2014; Revi et al. 2014). Moreover, as Caravani et al. (2014: 2) point out, "The scope of what constitutes an adaptation project is wide", making it difficult to establish whether donors are allocating funds in support of climate change adaptation, as opposed to other policy objectives. Although the UNFCCC provides a number of Articles (4.1, 4.4, 4.8 and 4.9) that identify the ways in which parties to the convention may be expected to support adaptation, the convention lacks a precise definition that can be operationalized through the UNFCCC (Burton 2004 [2009]).

According to UNEP (2016: 27), 90% of all bilateral funding (i.e. country to country) for climate change adaptation in 2014 went to non-OECD countries. Of this, an estimated 46% went to the Asia-Pacific, followed by sub-Saharan Africa (14%), Latin America and the Caribbean (12%) and South Asia (9%). Although this represents an improvement in the regional and geographical distribution of adaptation funding, international financing remains heavily skewed in favour of a relatively small number of recipients (Caravani et al. 2014; Shi 2016; UNEP 2016). In 2014, for instance, the top recipients of adaptation-related bilateral ODA were India, Myanmar, the Philippines and Vietnam (UNEP 2016: 27).

Questions have also been raised about the sources and modalities that have been put in place for financing multilateral climate adaptation. According to UNEP (2016: xiii), 84% (or US$21 billion) of all adaptation funding in 2014 came from international financial institutions (e.g. the World Bank, the regional development banks) in the form of "low-cost or market-rate project debt (53% and 26% of the total, respectively)". Correspondingly, the vast majority (55%) of adaptation projects were funded by the IFIs involved "hard" infrastructural investments in water and wastewater management (UNEP 2016: xiii). That so much of the international disbursement has been made on concessional and market-rate terms suggests a strong "cost-recovery" bias in favour of "hard" adaptation options (Sovacool 2011) that maximize a return on investment (UNEP 2016).

To summarize, the ability of cities to influence and exact benefits from the multilateral regime appears to be significantly constrained by a lack of resources and capacity on the part of many cities and a lack of funding and attention on the part of donors, national governments and the UNFCCC. Much like the international negotiations that gave them force, the UNFCCC and the Paris Climate Agreement are part of a Westphalian system whose principal aim is to uphold the jurisdictional and territorial sovereignty of individual nation-states. Correspondingly, most of the adaptation programming that has been supported under the UNFCCC has transpired on a country-by-country basis (Caravani et al. 2014; UNEP 2016). To the limited degree that cities and urban issues have been prioritized in multilateral programming, adaptation has largely been of the hard variety, focusing on large infrastructural investments.

4.3 Network Power

As noted in Chap. 3, transnational city-networks have come to play a critical role in building the capacity of cities for reducing their carbon footprint. Beyond or alongside the more "traditional" roles they have played in acting and advocating on behalf of cities is the intriguing notion that city-networks are also now "orchestrating" new forms of governance by establishing measures, metrics and compliance mechanisms that render cities observable and therefore governable in a wider social field of practice (Bulkeley 2010; Gordon and Acuto 2015; Acuto and Rayner 2016; Hickmann 2016; Castan Broto 2017; Gordon and Johnson 2017). One recent example of transnational city-networks collaborating on climate resilience is *Mayors Adapt*, an adaptation initiative that was launched in 2014 by the European Commission and the Covenant of Mayors "to engage cities in taking action to adapt to climate change".[10]

The implication is that city-networks are exerting new forms of power that entail establishing and reinforcing relations, expectations and obligations with member and non-member cities. But what do we know about this power, and how do does it manifest itself in the field of adaptation?

As was the case with mitigation (Chap. 3), city-network efforts to build the resilience of cities have also tended to cluster around larger cities, whose economic, political and diplomatic clout enables them to occupy pivotal niches within climate policy networks (Acuto 2013; Shi 2016). One example of this is the Rockefeller Foundation's *100 Resilient Cities* programme, whose "10% pledge" has (thus far) enlisted 28 cities to put aside 10% of their municipal budgets in the name of building resilience at the urban scale.[11] According to its Website, the *100 Resilient Cities* programme was founded at the end of 2013 and "is dedicated to helping cities around the world become

[10]http://www.covenantofmayors.eu/Adaptation.html. Last accessed 8 August 2017.

[11]http://www.100resilientcities.org/blog/entry/10percent-resilience-pledge#/-_/. Last accessed 30 May 2017.

more resilient to the physical, social and economic challenges that are a growing part of the twenty-first century".[12] Among the measures that it advocates for building urban resilience is a combination of "soft" (e.g. building social networks, preserving wetlands, creating a sense of collective identity and mutual support, fostering economic prosperity) and "hard" (e.g. building sea walls, investing in communication and mobility) adaptation strategies.[13]

However, as Shi (2016) has pointed out, the vast majority of cities participating in the *10% pledge* programme are located "in rich and middle income countries with Kigali, Rwanda, being the exception". Moreover, the measures being promoted have been criticized for advocating actions whose ability to build urban resilience is by no means guaranteed. One anonymous comment on the *10% pledge* Website, for instance, questioned the wisdom of promoting sea walls as a means of reducing flood vulnerability, suggesting instead that "floodplain reclamation coupled with better land use and stormwater management are significantly more climate resilient than building levee walls".[14]

Comments such as these point to the selective and ambiguous nature of climate adaptation, as well as the limited extent to which city-networks have (thus far) been able to expand beyond their traditional constituencies. If we look at the Global Covenant of Mayors for Climate and Energy, for instance, we can see that the vast majority of member cities are based in Europe and North America, which is unsurprising given that the Covenant was established (in 2016) as a result of an amalgamation of the Compact of Mayors and European Covenant of Mayors.[15] Similarly, the C40 Cities Leadership Group offers adaptation programming (including support for the Global Covenant), but here too the membership is largely confined to larger cities with greater capacity and influence.

[12]http://www.100resilientcities.org/about-us#/-_/. Last accessed 10 May 2017.

[13]http://www.100resilientcities.org/resilience#/-_/. Last accessed 30 May 2017.

[14]http://www.100resilientcities.org/blog/entry/10percent-resilience-pledge#/-_/. Last accessed 30 May 2017.

[15]http://www.globalcovenantofmayors.org/?welcome. Last accessed 31 May 2017.

Whether these disparities are the result of city-networks targeting wealthier cities or whether they are simply focusing on cities and networks with whom they already have pre-existing ties (cf. Lee 2015), observations have also been made that there are many vulnerable cities that have not engaged extensively in transnational city-networks, and that the composition of these networks is often limited to key decision-makers (e.g. mayors, senior city officials, academics, civil society and industry representatives) as opposed to the populations that are most vulnerable to the impacts of climate change (e.g. homeless people, slum dwellers, migrant labour) (Revi et al. 2014; Boyd et al. 2015; Gore 2015; Johnson et al. 2015).

As we have already seen in the case of urban climate mitigation (Chap. 3), the factors affecting city engagement in transnational networks can be usefully understood in terms of the opportunity and transaction costs of city leaders, officials and other stakeholders participating in international meetings and discussions, adopting transnational metrics and committing to network standards and objectives. For mitigation, the costs of engaging in transnational city-networks are largely about incorporating measures, metrics and standards that reduce the city's carbon footprint and justifying these measures in the eyes of local stakeholders and interests. For adaptation, the costs and benefits are in theory more closely aligned with local demand for clean water, safe housing, affordable energy, etc., but trade-offs exist between the costs and perceived benefits of changing as opposed to maintain maladaptive pathways (Kennedy 2011; Leichenko 2011; Sovacool 2011; Sovacool et al. 2015; Romero-Lankao et al. 2016; Castan Broto 2017).

On this point, considerable attention has been paid to the impact of city-networks on local policy processes and decisions, including the real and perceived costs of investing in adaptation measures at the urban scale (Anguelovski et al. 2014; Carmin et al. 2012; Boyd et al. 2015; Gore 2015; Sovacool et al. 2015; Shi 2016; Chu et al. 2015, 2017). As noted in Chap. 3, city-networks are often highly dependent upon external funding from national governments, international donors and the private sector, suggesting that their ability to engage in adaptation policy and programming that is autonomous from donors, national governments, private foundations, multinational corporations and the multilateral regime may be limited (Hickmann 2016;

Gordon and Johnson 2017). A related issue is that many cities simply lack the time, resources, contacts and capital to participate and engage in city-networks, highlighting the concern raised by Shi and colleagues that:

> … global adaptation networks may contribute to, rather than reduce, the inequitable distribution of adaptive capacities and resources across cities and nations. (Shi et al. 2016: 133)

Whether and to what extent cities will be able to engage in "global adaptation networks" will depend on many factors, including the wealth, size, sectoral specificity and geographical proximity of individual cities, as well as the particular rules, norms and relational aspects of city-networks. In what follows, I focus primarily on the latter, comparing three influential city-networks: (1) the Global Covenant of Mayors; (2) ICLEI; and (3) the C40.

4.3.1 The Global Covenant of Mayors

Just as it does for mitigation (Chap. 3), the Global Covenant offers a four-stage process in which cities are required to (1) make a commitment; (2) take inventory; (3) create measurable targets; and (4) establish an action plan. Box 4.3 describes the specific steps that cities are expected to take, but here it is worth noting that adaptation is considered "Step 2" of the four-stage process, reiterating the aforementioned bias in favour of mitigation.

Box 4.3 Building urban resilience: The Global Covenant's four-stage process[16]

Stage 1. Making a commitment

- Address the impacts of climate change
- Identify climate hazards
- Assess vulnerabilities
- Develop climate adaptation plans

[16] https://www.compactofmayors.org/resources/#tab-phase-1. Last accessed 31 May 2017.

Stage 2. Taking inventory

- Identify climate hazards
- Report in chosen platform

Stage 3. Creating measurable targets

- Assess climate vulnerabilities using Compact standard
- Report in chosen platform

Stage 4. Establishing an Action Plan

- Develop climate resilience plan demonstrating how city will deliver on its commitment to adapt to the impacts of climate change
- Report in chosen platform.

As the language in Box 4.4 suggests, the procedures that prospective member cities are expected to follow are far less detailed and established than is the case with mitigation (Chap. 3). For stages 2–4, for instance, cities are required to report in a "chosen platform", but—apart from the *City Climate Hazard Taxonomy*, a classification system that was developed in collaboration with ARUP and the C40 (see below)—all of the tools that are listed on the Compact/Covenant Website (i.e. CDP, carbonn and the GPC) are for mitigation only.

In fairness, the Global Covenant (including the Compact and the European Covenant of Mayors that preceded it) is a new initiative, and it is clear that orchestrating urban climate mitigation has taken a priority (for the time being) over adaptation. That said, it is difficult to avoid the observation that the heavy emphasis that has been placed on reducing emissions remains highly consistent with the strong mitigation agenda that has been advanced within the multilateral process (Burton [2004 (2009)]; Revi et al. 2014).

4.3.2 ICLEI

Taking a very different trajectory (that also includes collaborating with the Covenant of Mayors and the C40) is ICLEI, a transnational city-network that was originally "conceived in 1989 when 35 local government leaders from Canada and the USA met with a leading atmospheric scientist to discuss the depletion of the ozone layer".[17] As noted in Chap. 3, ICLEI has been actively involved in a variety of urban climate governance initiatives, including the *Cities for Climate Protection Programme*, the *Global Protocol for Community-Scale Greenhouse Gas Inventories* (the GPC), carbonn and, indeed, the Global Covenant of Mayors (Betsill and Bulkeley 2004; Hickmann 2016).

Although its principal focus has been on mitigation, ICLEI also offers a number of global and regional programmes that fall under the heading of urban adaptation. Arguably, one of the network's most visible policy platforms is *Resilient Cities, the Annual Global Forum on Urban Resilience and Adaptation.* Held every year in the German city of Bonn, the Forum brings together researchers, public officials, politicians, community leaders and business representatives "to discuss adaptation challenges facing urban environments around the globe".[18] As Hickmann (2016) has argued, the Global Forum has come to play an important role in sharing information, building capacity and articulating urban needs and interests, as have the literally hundreds of regional meetings and workshops that ICLEI organizes every year.

In contrast to other networks like *100 Resilient Cities* or the C40 (see below), ICLEI is a large and diverse network with a relatively decentralized governance structure. In addition to the World Secretariat (which is based in Bonn), ICLEI has 9 regional offices that are involved (to varying degrees) in governing the network. According to Hickmann (2016: 75),

> The supreme body of the ICLEI network is the Council that represents the network's overall membership. The Council is composed of several Regional Executive Committees representing different world regions. Each Regional Executive Committee appoints one member to the Global Executive Committee, which is the governing body of the network.

---

[17]http://archive.iclei.org/index.php?id=748. Last accessed 1 June 2017.

[18]http://resilient-cities.iclei.org/resilient-cities-hub-site/about-the-global-forum/. Last accessed 31 May 2017.

Because of this decentralized structure, ICLEI's adaptation programming tends to involve a wide array of local and regional partnerships, funded primarily by national governments, international donors and (in some cases) the private sector (Hickmann 2016). Alongside its global climate programming (carbonn, the *GPC* and the *Green Climate Cities Program*), ICLEI's regional offices have developed and/or collaborated on a large number of high-profile programmes, including the *Asian Cities Climate Change Research Network* (ACCCRN),[19] the *Future Resilience for African Cities and Land* (FRACTAL)[20] and *Building Adaptive and Resilient Communities* (a Canadian initiative, known as BARC).[21]

Compared with other city-networks, ICLEI has a long record of working with smaller, regional cities, businesses, national governments, donors and other city networks in building urban resilience, reinforcing its claim to be the "leading global network of over 1500 cities, towns and regions committed to building a sustainable future".[22] At the same time, its ability to extend beyond the pre-existing network is constrained by a number of path dependencies, as well as a lack of resources. To take but one example, the Website for ICLEI's "South American" regional office is written only in Portuguese, suggesting a major constraint on its ability to engage with regional partners whose first language is Spanish, English or French. As Hickmann (2016: 77) has pointed out, blind spots such as these are not necessarily the result of an intentional bias on the part of the network, as opposed to a lack of stable funding and resources:

> ...the ICLEI network faces considerable budgetary constraints and lacks the necessary capacities to launch effective programs and projects to address the problem of climate change.

[19]http://southasia.iclei.org/our-activities/our-agendas/resilient-city/asian-cities-climate-change-resilience-network-acccrn.html. Last accessed 31 May 2017.

[20]http://www.iclei.org/activities/agendas/resilient-city.html. Last accessed 31 May 2017.

[21]http://www.icleicanada.org/programs/adaptation. Last accessed 31 May 2017.

[22]http://www.iclei.org/about/who-is-iclei.html. Last accessed 1 June 2017.

According to its triennial 2012 Annual Report (ICLEI 2012), ICLEI's global budget for 2011 was 12.5 million euros.[23] Outside of membership fees (12%), "fees for services" (e.g. conferences—20%) and local government contributions (14%), the largest sources of revenue for the network were private foundations and associations (11%), international agencies (11%) and national, sub-national and supranational governments, including the EU (29%).

In this respect, ICLEI's experience is comparable to many international NGOs, whose mandate has expanded, thereby necessitating new forms of collaboration and investment from donors, national governments and the private sector.

4.3.3 The C40

Arguably the most important (and ambitious) example of corporate-network collaboration in the field of urban climate governance has been the C40. As noted in Chap. 3, the C40's work on urban climate governance has involved extensive collaboration with corporations (like ARUP), private foundations (such as Bloomberg Philanthropies and the Rockefeller Foundation) and international agencies (such as UCLG, UN-Habitat, the World Bank and the UNFCCC). The implications of corporate involvement in network governance we explore in due course, but here we first describe some of the C40's principal areas of work on urban climate adaptation and resilience.

In contrast to ICLEI, the C40 is a much smaller network of large and influential cities, "representing over 600 million people and one quarter of the global economy".[24] As noted in Chap. 3, the network tends to be confined to wealthier cities, whose influence and capacity are assumed to provide a more effective means of developing, testing and disseminating climate governance models and agendas (Gordon and Acuto 2015). One of its flagship adaptation initiatives is *the Connecting Delta Cities Programme,* a

---

[23]http://archive.iclei.org/fileadmin/user_upload/documents/Global/governance/ICLEI-Corporate_Report2012-final-www.pdf. Last accessed 1 June 2017.

[24]http://www.c40.org/about. Last accessed 9 December 2016.

network of 13 cities, whose principal aim is to "develop a network of delta cities that are active in the field of climate change related spatial development, water management, and adaptation, in order to exchange knowledge on climate adaptation and share best practices that can support cities in developing their adaptation strategies".[25] Another is the *Climate Risk Adaptation Framework and Taxonomy*—or *CRAFT*, which the C40 Website describes as "a standardized reporting framework that enables cities to perform robust and consistent reporting of local climate hazards and impacts, risk and vulnerability assessment, and adaptation planning and implementation as part of their compliance with the Compact of Mayors".[26]

In comparing these two programmes, the CDC is largely about sharing information and best practices that might lead to "innovative adaptation strategies", but thus far the network has been limited to a small number of member cities. Apart from Jakarta, Rio de Janeiro and Ho Chi Minh City, all of the CDC cities are in industrialized countries; Rio is the only city in South America; there are no African cities.

CRAFT, on the other hand, is more ambitious, providing a "reporting standard" that in the words of the C40, is:

> ...integrated into the CDP Cities and carbonn Climate Registry reporting platforms, which allows cities to share the status of their climate adaptation planning efforts. It establishes a platform for city officials to report the unique conditions their cities face and their planning responses.[27]

Box 4.4 Joining the C40's *CRAFT* initiative[28]

Profile the City: This module collects city data on general characteristics that influence city resilience and adaptation planning. This information can support shared learning and inform adaptation planning at the local level.

---

[25]http://www.deltacities.com/about-c40-and-cdc. Last accessed 1 June 2017.

[26]http://c40-production-images.s3.amazonaws.com/other_uploads/images/445_C40_CRAFT_v11.original.pdf?1453129528.

[27]http://c40-production-images.s3.amazonaws.com/other_uploads/images/445_C40_CRAFT_v11.original.pdf?1453129528.

[28]http://c40-production-images.s3.amazonaws.com/other_uploads/images/445_C40_CRAFT_v11.original.pdf?1453129528.

Understand the Problem: This module asks the city to report on their vulnerability assessment process, the climate risk and vulnerability faced now and in the future, and the underlying factors within a city that can enhance or challenge a city's ability to adapt.

Plan, Respond and Monitor: This module is focused on the climate adaptation planning process undertaken by cities and how cities evaluate the outcomes of their adaptation efforts. Reporting fields relate to adaptation planning, adaptation goals and actions, and adaptation barriers and opportunities.

As can be seen in Box 4.4, *CRAFT* involves some of the very same reporting and assessment processes that have been used by other city networks. What makes it different, though, and what connects it to the following section on corporate power, is the way in which it classifies, assesses and understands urban risk and vulnerability to climate change. In the same way that corporations like ARUP have orchestrated and incorporated cities into new social fields of observation, classification and assessment, the C40 and *CRAFT* in particular have created a reporting and planning protocol that renders cities observable and governable in the field of urban climate governance.

4.4 Corporate Power

From the preceding, we can infer that cities and city-networks have become increasingly dependent upon collaborations with multinational corporations and private foundations, such as Siemens, Arup, Bloomberg Philanthropies and the Rockefeller Foundation (Gordon and Acuto 2015; Acuto and Rayner 2016; Hickmann 2016; Gordon and Johnson 2017). However, the implications of corporate involvement in urban adaptation planning and governance remain poorly understood, reflecting the speed and scale at which city-network-corporate alliances have evolved over the last ten to fifteen years (Davidson and Gleeson 2015; Gordon and Acuto 2015; Gordon 2016; Acuto and Rayner 2016; Gordon and Johnson 2017).

Cast in their most positive light, private-sector collaborations provide what appears to be an essential source of finance and investment that can be used in developing low-carbon, climate-resilient pathways (Arup-C40 2015a, b; 2016). However, securing access to capital entails adopting and adhering to industry norms and standards, suggesting that the ability of cities and city-networks to project power and influence is to a certain degree constrained by their connections (through institutions, interests and metrics) with private-sector donors and investors (Bouteligier 2015; Davidson and Gleeson 2015; UNEP 2016; Gordon and Johnson 2017). In the words of UNEP (2016: xiv):

> Outside of a purely adaptation-related context, private sector contributions – from foreign direct investment, private debt, remittances and official development assistance – make-up the largest components of financial inflows to developing countries. The distribution of flows is uneven, with least developed countries struggling to attract significant volumes of private debt or equity outside resource sectors.

A related concern is that the models being used to frame urban climate adaptation are rooted in a particular understanding of risk and vulnerability that fails to capture the "hidden" forms of vulnerability that are only rarely (if ever) reported in official documentation (Satterthwaite and Bartlett 2017). As Davidson and Gleeson (2015) have argued, the "strategic urbanism" that Arup and the C40 have adopted in their policy and programming appears strongly pre-disposed towards an "auditing culture" that disseminates, verifies, certifies and institutionalizes a prescribed range of city actions on climate change (cf. Gordon and Johnson 2017). Whether these actions lead to substantive improvements in urban climate resilience remains a matter of some debate (Davidson and Gleeson 2015; Widerberg and Stripple 2016; Gordon and Johnson 2017; Hodson et al. 2017).

Particularly important in the context of urban climate adaptation is the extent to which corporate models and approaches to climate governance are disseminating norms, standards and metrics that effectively redefine what it means to be vulnerable. In what follows, we consider two interrelated forms of corporate climate governance: (1) management consulting and (2) urban climate insurance.

4.4.1 Adopting or Adapting? The Promise (and Peril) of Management Consulting

Management consulting is a term that has been used to describe a wide range of professional services aimed at assisting public, private and not-for-profit institutions in identifying and achieving their core goals and objectives. In a context of economic globalization and public sector restructuring, the sector underwent considerable growth in the 1990s and 2000s (Christensen et al. 2013). Although the field is varied, a central dimension of management consulting entails identifying the internal/organizational and external/environmental factors that affect the ability of organizations to meet their core goals and objectives. In the words of one American consulting firm,

> Business strategy consulting is about facts and common sense much more than vision or dreams. Facts make decisions easier. Dreams without facts make them tougher.[29]

Arguably the most pervasive form of management consulting in city and city-network climate governance can be found in the corporate policy and planning documents that have been used to frame the aims and agendas of city-networks (e.g. Arup-C40 2015a, b; 2016; Arup-Rockefeller 2016). Underlying the *CRAFT* methodology, for instance, is a taxonomy of terms and metrics that are used to ensure that the information being provided by member cities can be systematically registered and recorded with network reporting platforms, such as CDP and carbonn.[30] According to its Website, *CRAFT* provides:

- A framework for cities to perform robust and consistent reporting of climate hazards and associated adaptation planning and implementation that is required by the Compact of Mayors;

[29]http://www.bain.com/offices/brussels/en_us/careers/top-management-consulting.aspx. Last accessed 23 June 2017.

[30]http://c40-production-images.s3.amazonaws.com/other_uploads/images/445_C40_CRAFT_v11.original.pdf?1453129528.

- A means to monitor and evaluate adaptation planning progress to help cities improve adaptation efforts by enhancing knowledge of best practices;
- A means for cities to identify priorities and target advocacy for climate adaptation resources;
- The data to improve the ability for cities and their partners to identify peers and aspirational examples to help inform their own adaptation planning process and implementation.

An important aspect of the *CRAFT* methodology is its classification of hazards into "low", "medium" and "high risk", which it then uses to match cities with "similar hazard risk profiles". According to the Website, profile matching is intended to maximize policy learning and information sharing among cities, but it also provides an important means of developing a database that can be used to identify systemic patterns of vulnerability and risk, thereby facilitating the development of financial instruments that may be used in pooling and transferring urban risk (see below).

Similar in purpose (if not in scope) is the *City Resilience Index*, a comprehensive list of monitoring and assessment indicators that were developed by Arup and the Rockefeller Foundation as a means of diagnosing "strengths and weaknesses and measure relative performance over time". According to its Website,[31]

> The Index has been designed to enable cities to measure and monitor the multiple factors that contribute to their resilience. It's (*sic*) primary purpose is to diagnose strengths and weaknesses and measure relative performance over time. This provides a holistic articulation of city resilience, structured around four dimensions, 12 goals and 52 indicators that are critical for the resilience of our cities (Arup-Rockefeller 2016: 2).[32]

Underlying the 12 goals and 52 indicators are four thematic dimensions that are used to frame city reporting:

[31]http://publications.arup.com/publications/c/city_resilience_index. Last accessed 1 June 2017.

[32]http://publications.arup.com/publications/c/city_resilience_index. Last accessed 1 June 2017.

- *Health and well-being*, ensuring the health and well-being of everyone living and working in the city;
- *Economy and society*, the social and financial systems that enable urban populations to live peacefully and act collectively;
- *Infrastructure and environment*, man-made and natural systems that provide critical services, protect and connect urban citizens; and
- *Leadership and strategy*, the need for informed, inclusive, integrated and iterative decision-making in our cities.

Much like the GPC (Chap. 3), *CRAFT* and the *City Resilience Index* provide a systematic means of reporting and evaluating individual city actions, thereby generating a body of evidence that may be used for "baselining" and "agenda setting" the needs and priorities of individual member cities. However, questions can be raised about the spatial and socio-economic dimensions of the urban adaptation being proposed in this context (Shi et al. 2016).

First, both *CRAFT* and the *Cities Resilience Index* are based on what appears to be a limited number of relatively wealthy cities, whose populations are arguably unrepresentative of the larger number of secondary and low-income cities facing wide ranging vulnerabilities to climate change (Shi et al. 2016). Moreover, the time and scope for the data gathering that went into both sets of indicators appear insufficient for documenting the "hidden vulnerabilities" of poor and marginal populations that are arguably most vulnerable to the impacts of climate change (Shi et al. 2016; Satterthwaite and Bartlett 2017).

Second, and related, questions can be raised about the validity of the indicators being used to measure city resilience. Although they are both comprehensive in scope, *CRAFT* and the *Cities Resilience Index* are in the final analysis cross-sectional snapshots that equate urban vulnerability with "an end point of a linear process", as opposed to a dynamic process that captures "how and why particular cities or populations are more vulnerable – or more able to cope or adapt than others" (Romero-Lankao and Qin 2011: 143–145). The result is an amalgam of urban actions and conditions that provide at best an undifferentiated proxy for urban climate resilience and vulnerability (Davidson and Gleeson 2015).

Third, there is a noticeable lack of detail on the kinds of policies that may be necessary for reducing the resilience gaps that are being documented in these framing documents. If we take, for instance, the first goal of achieving *minimal human vulnerability, the Cities Resilience Index* lists: 1.1 safe and affordable housing; 1.2 adequate affordable energy supply; 1.3 inclusive access to safe drinking water; 1.4 effective sanitation; and 1.5 sufficient affordable food supply as indicators of what constitutes minimal human vulnerability at the urban scale. Similarly, the seventh goal of *reduced exposure and fragility* lists: 7.1 comprehensive hazard and exposure mapping; 7.2 appropriate codes, standards and enforcement; 7.3 effectively managed protective ecosystems; and 7.4 robust protective infrastructure.

Although the indicators are intended to measure and assess urban resilience, the critical point here is that they are also embedded in a wider range of conceptual and political assumptions that pre-define what constitutes "safe and affordable housing", "adequate affordable energy", "appropriate codes, standards and enforcement" and "robust effective infrastructure". Notwithstanding the claim (made, for instance, in the Introduction to the City Resilience Index; Arup-Rockefeller 2016) that systematic indicators can be tailored to meet the specific needs and priorities of individual cities, the lack of attention to specific policy approaches (e.g. using rent controls and new housing construction to provide "safe and affordable housing") has the effect of depoliticizing the political struggles and trade-offs that typically influence urban policy processes and outcomes. Moreover, the lack of attention to specific policy approaches and agendas facilitates the amalgamation of seemingly redistributive policy issues, such as urban safety and social housing, to market-based solutions, such as carbon pricing and public–private partnerships in the delivery of urban infrastructure (Davidson and Gleeson 2015).

Fourth, questions can be raised about the range of assets, livelihoods, populations and interests that are being represented in these methodologies and repositories of information (Davidson and Gleeson 2015; Shi et al. 2016). Although one cannot go too far in linking the assessment tools that *CRAFT* and the *Cities Resilience Index* are using to measure urban resilience with a wider "corporate" agenda, it is difficult to avoid observing that companies like Siemens and Arup are also major providers of urban planning services and, in theory at least, they stand to benefit from selling these services to cities and urban enclaves that can afford to pay. Alongside the work, it has been doing on urban resilience and climate change, for instance, Arup also offers services on "smart cities" and urban renewal.[33]

Box 4.5: How smart is your city? Urban adaptation as smart city governance

"Smart city" governance is a term that has been used to describe the integration of information and communication technology with the delivery of urban services, such as transportation, education, healthcare, waste disposal, water treatment, law enforcement and urban energy. Within the context of urban climate governance, it is seen to offer an important means of reducing emissions and urban vulnerability by developing more effective/efficient ways of integrating resource waste and consumption at the urban scale. Recent efforts to include smart city principles into urban planning include the cities of Stockholm, Barcelona, Singapore and Madrid. The Government of India also introduced its Smart Cities Mission in 2015. Although they are now being promoted by planners, politicians and corporations around the world (including Siemens and Arup), smart city initiatives have been criticized for promoting a model of technocratic urban development that prioritizes surveillance and meta-management at the expense of personal freedom and mobility (Sennett 2012; Hodson et al. 2017). It has also been argued that smart city technologies will be expensive, raising questions about the viability and scalability of adopting these models in the Global South (Roberts 2016).

Questions about the representativeness, validity, specificity and intended beneficiaries of management consulting resilience indicators highlight the notion that corporate approaches to assessing, pooling and governing urban risk may not reflect the spatial and socio-economic disparities that exacerbate urban vulnerability to climate change (Shi et al. 2016).

---

[33]http://www.arup.com/markets/cities. Last accessed 23 June 2017.

4.4.2 Closing the "Protection Gap": Cities, Insurance and Climate Risk

The insurance industry is often looked upon as a key player in the global fight against climate change.[34] On the mitigation front, it took an early stand in favour of curbing global emissions. With adaptation, it has also argued consistently in favour of improving local codes and by-laws for reducing losses due to windstorms, floods and other climatic hazards. At the same time, the industry's approach to pooling and transferring risk is based on a model that typically favours larger and wealthier clients and assets, as opposed to informal sectors and low-income populations in the developing world.[35]

But this may be changing. In recent years, increasing attention has been paid to the "protection gap" that has emerged between the *insured losses* that insurance companies have been required to pay out to their clients and the *economic losses* that have been incurred on the part of businesses, homeowners, governments and cities lacking formal coverage. According to Swiss Re, one of the world's largest insurance companies, economic losses in 2016 exceeded $175 billion, which represents an increase of $81 billion from 2015.[36] Over the same period, insured losses from natural disasters and "man-made loss events" were $54 billion, double the claims that were made in 2015, representing a protection gap of $121 billion.

Although 2016 was an unusually destructive year for the insurance industry (including major earthquakes in Italy, Indonesia and Ecuador, windstorms in Haiti, and a number of highly devastating floods in the USA), the gap between the "rate of growth of economic losses has outpaced the growth of insured losses over the last 25 years",[37] reflecting the effects of urbanization and wealth accumulation outside of

---

[34]http://citiscope.org/story/2017/sustainable-insurance-seven-takeaways-first-ever-summit-insurers-and-city-leaders?utm_source=Citiscope&utm_campaign=4bf21fa118-Mailchimp_2017_05_12&utm_medium=email&utm_term=0_ce992dbfef-4bf21fa118-118068273.

[35]https://www.reinsurancene.ws/protection-gap-not-just-emerging-market-issue-swiss-re/.

[36]https://www.reinsurancene.ws/2016-global-insured-losses-54-billion-swiss-re/.

[37]https://www.reinsurancene.ws/protection-gap-not-just-emerging-market-issue-swiss-re/.

Western Europe and North America. According to Swiss Re, the Asian economies accounted for over two-thirds of the global disasters in 2016, causing $60 billion in economic losses.

However, emerging economies remain under-insured, raising concerns about the protection gap that exists between the exposure and coverage of urban assets and infrastructure. From the insurance industry's perspective, the challenge of closing the gap has entailed extending new forms of coverage and introducing measures to curb risky behaviour. Recent discussions within the industry have focussed on the need to develop better instruments for pooling risk (see below) and a more concerted effort on the part of private insurers and public institutions to improve the data modelling and assessment that is used to decide local by-laws and zoning regulations.[38] Particularly important in this regard is the need to reduce pre-disaster risk, as opposed to mitigating the effects of disasters after they occur.

In theory, the insurance industry is poised to scale up new approaches to reducing urban climate risk. The industry had more than $30 trillion (USD) of protected assets in 2016, and it clearly has an interest in expanding its coverage to new clients (and possibly new hazards) in emerging economies.[39] In practice, its ability to extend coverage is itself embedded in risk, highlighting the role of risk pooling, risk transfer (e.g. reinsurance) and other financial instruments.

One instrument that is now being discussed in the field of urban climate governance is *parametric insurance*. Similar to index-based insurance (Heltberg et al. 2009), parametric insurance uses a quantifiable or a parametric trigger as a means of establishing whether and when a payout needs to be made.[40] Examples of parametric triggers include rainfall, wind speeds and salinity levels, but the key point here is that the

[38]http://citiscope.org/story/2017/sustainable-insurance-seven-takeaways-first-ever-summit-insurers-and-city-leaders?utm_source=Citiscope&utm_campaign=4bf21fa118-Mailchimp_2017_05_12&utm_medium=email&utm_term=0_ce992dbfef-4bf21fa118-118068273.

[39]http://citiscope.org/story/2017/sustainable-insurance-seven-takeaways-first-ever-summit-insurers-and-city-leaders?utm_source=Citiscope&utm_campaign=4bf21fa118-Mailchimp_2017_05_12&utm_medium=email&utm_term=0_ce992dbfef-4bf21fa118-118068273.

[40]http://www.artemis.bm/library/what-is-parametric-insurance.html.

payment occurs only after the trigger occurs, thereby eliminating the need to verify individual claims and losses. A second form of insurance that appears to have great potential for cities and urban populations is the use of *catastrophic bonds*. Catastrophic bonds are high-yielding debt instruments that are used to raise money in the event of a catastrophic disaster, such as a hurricane or an earthquake.[41] A key feature of catastrophic bonds is that they indemnify issuers (i.e. an insurance or reinsurance company, or a local authority) from having to pay the coverage, either deferring payments or forgiving them entirely. From the issuer's point of view, the attraction is that it covers the risk of a catastrophic payout, but it still ensures that insured parties remain insured.

Beyond the idea of developing new instruments that can be used for sharing and transferring climate risk, the insurance industry stands to make an important difference to the ways in which local authorities decide land and resource use at the urban scale. Critical here is the challenge of restricting and even relocating homes, buildings, roads and infrastructure that are situated in areas that are currently—or projected to be—vulnerable to climate change.[42] However, as noted earlier, the factors affecting people's decisions to live and work in areas that are known or deemed to be "unsafe" are multiple and varied, highlighting the role of poverty, differential wage rates and affordability in processes of rural–urban migration and urban settlement (Broto Castan 2017; Satterthwaite and Bartlett 2017). Moreover, the capacity of local authorities to regulate processes of land use, rural–urban migration and urban (re) settlement is highly uneven, suggesting the need for better data (and better data management) that may be used in governing urban climate risk (Chap. 5).

In short, the insurance industry stands to make an important contribution to reducing urban climate risk, but historically its coverage has focused on larger, wealthier assets, suggesting that its ability to extend

---

[41] http://www.investopedia.com/terms/c/catastrophebond.asp.

[42] http://www.theglobeandmail.com/opinion/editorials/the-floodwaters-are-rising-in-canada-when-will-ottawa-finally-deal-with-it/article34969402/?reqid=fe5d6b86-4aaa-4968-90a2-df8aaeb7a458. Last accessed 13 May 2017.

into smaller, poorer urban markets will be limited. That said, the convergence of corporate and city-network governance constitutes an important form of performative power that has important bearing on the ways in which cities are now repositioning themselves in the context of climate change.

4.5 Performative Power

One of the most striking aspects of contemporary climate governance is the extent to which transnational actors and networks have established their own rules, norms and procedures for reporting and evaluating climate policy actions and commitments at different scales (Betsill et al. 2015; Bulkeley et al. 2015; Bäckstrand and Kuyper 2017; Bäckstrand et al. 2017; van der Ven et al. 2017; Gordon and Johnson 2017). As noted in Chap. 3, the act of reporting, demonstrating and committing to transnational norms and objectives confers its own kind of performative power that renders cities and city-networks observable and therefore comparable in particular fields of governance. Whether a city is deemed a financial or environmental risk, for instance, has enormous bearing on its ability to attract loans, investment and insurance that may conceivably build urban resilience to climate change.

However, making sense of the power that cities have been able to claim and obtain in these instances poses challenges, not least the challenge of differentiating and determining where multilateral or network power ends and where corporate power begins. One basic form of power entails ranking—or being ranked—in relation to other cities. In recent years, increasing attention has been paid to the idea of ranking cities on a continuum of vulnerability to climate change (e.g. McGranahan et al. 2007; Hallegatte et al. 2013; *100 Resilient Cities*[43]). Explicit in these evaluations is a set of quantifiable criteria (e.g. population size, exposure to flooding, sea level rise and water-borne disease)

[43]http://www.100resilientcities.org/blog/entry/10percent-resilience-pledge#/-_/. Last accessed 30 May 2017.

that may be used in assigning a value to the relative vulnerability of cities and city regions. However, assessments are largely made on the basis of external standards, suggesting that cities themselves have little or no capacity to influence their place within the ranking.

A second and somewhat less tangible type of performative power entails mobilizing the symbols, images and narratives that may be used in justifying external support in the form of funding, special treatment or international assistance. Compared to an external ranking exercise, power of this kind assumes greater agency on the part of cities and city leaders to re-position themselves in relation to multilateral, network and corporate governance arrangements. In the age of social media, examples of this kind of performative power have become increasingly common, including mayors and city leaders declaring a central role for themselves in international climate negotiations or small island developing states (and even cities within small island developing states) using their "special" status of being "on the frontline" of climate change to justify additional funding, special recognition and loss and damage (Johnson 2017).

A third and critical form of performative power entails linking and integrating a city's risk and vulnerability assessments into multilateral, network and corporate fields of governance. In contrast to an external ranking exercise, power of this kind entails contributing and subjecting oneself to a set of standards that in and of itself confers standing and entitlement to external resources. In Gordon and Johnson's (2017: 706) words:

> The logic of orchestration in this case relies on a combination of instrumentality and structural inescapability. Cities may acquiesce to being orchestrated as either a means of, ironically, differentiating themselves from one another (to render oneself comparable is to establish the conditions on which to achieve status versus others). They may come to see the adoption of particular standards, or of practices of quantification and disclosure, as portals through which to access material or epistemic resources, as is the case for gaining access to a World Bank direct funding window for cities, for example.

Particularly important in this regard is the idea that subjecting oneself to an external set of norms and criteria (1) builds trust and predictability, thereby facilitating cooperation; and (2) provides a means by which orchestrators (e.g. the World Bank, the Green Climate Fund) and intermediaries (e.g. ICLEI, the C40) may manage and control the flow of information, finance, resources that may be used in governing urban climate risk. Examples here might include the *Cities Climate Finance Alliance*,[44] the *GEF Sustainable Cities Integrated Approach* initiative[45] and the World Bank's *City Creditworthiness Initiative*.[46] Of these three, the GEF initiative is particularly interesting in the sense that it seeks to place "high-risk bets on promising new technologies or approaches" that will emerge as "game changers" and have a "beacon effect" of spurring adoption elsewhere. In contrast, the World Bank's initiative seeks to link cities with potential donors and investors by improving their creditworthiness, including "the legal and regulatory, institutional, and policy framework for responsible sub-national borrowing".

In short, reporting and subjecting oneself to a common set of standards and criteria can facilitate access to other forms of loans and investment that are indispensable for financing long-term infrastructural development. However, entering transnational reporting exercises also creates its own forms of risk that may in and of themselves undermine the ability of cities to act autonomously or effectively on climate change. One relates to the reputational and financial risks of reporting economic, financial and environmental data for the purposes of investment. By providing data that facilitate the allocation of finance, credit and investment, cities are also exposing themselves to the possibility of receiving an unfavourable assessment in the form of downgraded bond ratings, higher interest rates or capital flight. For this reason, cities and city-networks (including the C40) are often very reluctant to share their financial data publicly.[47]

---

[44]http://www.citiesclimatefinance.org/. Last accessed 26 June 2017.

[45]https://www.thegef.org/topics/sustainable-cities. Last accessed 26 June 2017.

[46]http://www.worldbank.org/en/topic/urbandevelopment/brief/city-creditworthiness-initiative. Last accessed 26 June 2017.

[47]To date, the C40's financial reporting has been minimal. ICLEI releases a financial report only once every three years.

A second issue relates to the standards and indicators being used to measure and manage what constitutes acceptable climate risk. As Ulrich Beck (1992: 19) has argued, the social production of wealth in modern society is "systematically accompanied by" the social production of risk. Inherent in the methodologies and repositories of information being used to document and compare urban climate vulnerability is an assumption that urban adaptation is primarily a technical and managerial problem that can be solved through the application of finance, technological innovation and expertise. However, establishing what constitutes "safe and affordable housing", "adequate affordable energy supply", "inclusive access to safe drinking water" and other resilience indicators is itself a political process that prioritizes certain forms of urbanism at the expense of others (Davidson and Gleeson 2015; Sovacool et al. 2015; Broto Castan 2017).

Questions about politics, priorities and social exclusion raise a third point about the ways in which the standardization of risk and vulnerability affects the rigidity, vulnerability and integrity of an increasingly interconnected urban planetary system (Brenner and Theodore 2002; Brenner 2013; Brenner and Schmid 2013; Sassen 2015). Theories of complexity and resilience suggest that systems become increasingly rigid and vulnerable as they become increasingly centralized and interconnected (Holling 2001; Perrow 2011). In ecological systems, rigidities manifest themselves when a particular species becomes dominant, thereby preventing other competitors from utilizing resources within the system. Within human systems, they may take the form of rules and path dependencies that sustain existing structures, but at the same time make it less able to adapt and change in response to new pressures and events. Complex systems are ones in which interactions are multiple, unpredictable and generally unintended. Multiple feedback mechanisms, interconnected sub-systems, multiple and interacting controls, indirectly obtainable information and an incomplete understanding of the system make for high levels of unpredictability, which heighten the possibility of "cascading effects", events in which failures in one part of the system lead to unpredictable and uncontrollable failures in other parts of the system.

For cities and city-networks, system failures (e.g. blackouts, food shortages, financial crises) often manifest themselves when critical flows of resources, finances and energy are disrupted by system-wide dynamics (e.g. currency speculation, inflation) or ad hoc shocks (e.g. heat waves, droughts, etc.) (Perrow 1984, 2011; Kennedy 2011; Leichenko 2011; Romero Lankao et al. 2016). Here, the ability of cities to recover and adapt to system failures becomes highly dependent upon their ability to *de-couple* from system processes that have the effect of overwhelming local capacities (e.g. electricity grids dealing with catastrophic load demands), to *re-couple* with pre-existing systems and/or to *replace* old systems with new ones entirely. However, as we have seen in the previous two chapters, cities have become increasingly interconnected through processes of urbanization, policy diffusion and orchestration, reflecting the interactive nature of cities working with networks and multinational corporations in the multilateral climate regime (Gordon and Johnson 2017). At the same time, processes of urbanization, policy diffusion and orchestration have been highly uneven, suggesting that some cities and some urban populations have been able to benefit from these processes, while others have not (Anguelovski et al. 2014; Carmin et al. 2012; Sovacool et al. 2015; Shi 2016; Chu et al. 2015, 2017).

The implication then is that some cities have become more interdependent and possibly more rigid, thereby heightening the possibility of "cascading effects", in which failures in one part of the system lead to unpredictable and uncontrollable failures in other parts of the system. At the same time, the capacity of cities and vulnerable urban populations to adapt and react to extreme weather events is unevenly divided, highlighting critical questions about the ways in which and extent to which adaptation policies are in fact alleviating or exacerbating vulnerability to climate change.

A final issue concerns the distributional consequences of governing urban climate vulnerability as a form of risk. On this last point, Benjamin Sovacool and colleagues (2015: 616) have argued that understanding and evaluating climate change adaptation entails recognizing the "political, economic, and social trade-offs that can occur within or between adaptation projects". At the heart of their analysis is a typology

(Sovacool et al. 2015: 616–617) that differentiates four kinds of trade-offs that occur in a context of climate adaptation:

- *Enclosure*, which refers to "the process that transfers a public asset into private hands, or extends the role of a private actor into the public sphere";
- *Exclusion*, "by which adaptation projects limit access to resources or marginalize a particular group of stakeholders in the decision-making process";
- *Encroachment*, which "occurs when adaptation interventions intrude or infringe upon protected areas, national parks, and wildlife reserves, or interfere with the healthy functioning of an ecosystem"; and
- *Entrenchment*, which is "the process by which adaptation projects aggravate political, socio-economic, or cultural inequalities and the disempowerment of disadvantaged groups".

Taken together, processes of entrenchment intensify the inequities created by enclosure, exclusion and encroachment "by favouring concentration of wealth within a community, or by hurting vulnerable members of a community, such as artisanal fishers, indigenous groups, or ethnic minorities" (Sovacool et al. 2015: 617). The implication here is that urban risk governance is dependent upon the dispossession and displacement of populations that are most exposed to the impacts of climate change (e.g. homeless people, slum dwellers, migrant labour) (cf. Roy et al. 2013; Boyd et al. 2015).

From a policy perspective, important questions can therefore be raised about the challenge of protecting or incorporating marginal populations into local and transnational decision-making networks and processes (Archer et al. 2014; Boyd et al. 2015; Carmin et al. 2012; Chu et al. 2015; Forsyth 2013; Gore 2015; Haque et al. 2014; Roy et al. 2013). From an analytical standpoint, more fundamental questions can be raised about the processes of dispossession, displacement and resistance that underlie the orchestration of transnational urban climate governance (Brenner and Theodore 2002; Brenner 2013; Brenner and Schmid 2013; Harvey 2006; McCann 2011).

4.6 Concluding Remarks

From the preceding, we can infer that cities are entering and establishing a wide range of multilateral, network-governed and corporate arrangements and procedures that have important bearing on their ability to build resilience to climate change (e.g. NAZCA, CRAFT and the Cities Resilience Index). At the same time, the power and autonomy of cities remain highly dependent upon their ability to negotiate favourable terms with city-networks, corporations and nation-states. Within the multilateral process, cities have received what appears to be a tiny share of the international funding that has thus far been allocated for adaptation. Moreover, adaptation has received only a tiny share of multilateral climate finance.

Although the Green Climate Fund is still in its infancy, the relative lack of attention to urban adaptation suggests a policy bias that has manifested itself not only in the global adaptation regime (Shi et al. 2016), but also among city-networks themselves. According to ICLEI (2015), the vast majority of actions being documented by cities and city-networks have been directed towards mitigation.

Whether cities and city-networks will make a difference in terms of reducing vulnerability to climate change will depend on the ability of municipalities, businesses, epistemic communities and civil society organizations to invest in infrastructure and institutions that can be replicated and maintained in the face of future social and environmental stress (Atkins 2012; Leichenko 2011; World Bank 2010a). But it will also depend on the ability of cities to support policy initiatives that work with a wider range of state and non-state actors whose interests, actions and institutions have important bearing on the ability to engage in mitigation and adaptation at wider scales of interaction.

On this point, the involvement of private foundations and corporations represents something of a paradox for cities and city leaders. On the one hand, it provides the capital and expertise that are absolutely essential for building urban resilience. On the other, it locks cities (and city-networks) into powerful path dependencies (based primarily on hard technical solutions) that run the risk of exacerbating urban vulnerability to climate change. Particularly concerning is the possibility that adaptation and risk governance processes dispossess and displace populations that are already highly vulnerable to climate change.

Coming back to the distinction between adaptive and maladaptive development pathways, the evidence and argument presented in this chapter raise important questions about (1) whether and to what extent cities and city-networks are able to act autonomously and effectively on climate change; (2) whether the processes and changes being documented in the chapter are making cities and urban populations more or less vulnerable to climate change.

References

Acuto, Michele. 2013. *Global cities, governance and diplomacy: The urban link*. New York and London: Routledge.

Acuto, M., and S. Rayner. 2016. City-networks: Breaking gridlocks or forging (new) lock-ins? *International Affairs* 92 (5): 1147–1166.

Adger, W.N. 2006. Vulnerability. *Global Environmental Change* 16 (3): 268–281.

Anguelovski, I., E. Chu, and J. Carmin 2014. Variations in approaches to urban climate adaptation: Experiences and experimentation from the global South. *Global Environmental Change* 27: 156–167.

Archer, Diane, Florencia Almansi, Michael DiGregorio, Debra Roberts, Divya Sharma, and Denia Syam. 2014. Moving towards inclusive urban adaptation: Approaches to integrating community-based adaptation to climate change at city and national scale. *Climate and Development* 6 (4): 345–356. doi:10.1080/17565529.2014.918868.

Armitage, D., M. Marschke, and R. Plummer. 2008. Adaptive co-management and the paradox of learning. *Global environmental change* 18 (1): 86–98.

Arup-C40. 2015a. *Climate action in Megacities 3.0*. http://www.cam3.c40.org/images/C40ClimateActionInMegacities3.pdf. Last accessed 14 June 2017.

Arup-C40. 2015b. *Potential for climate action*. http://publications.arup.com/publications/p/potential_for_climate_action. Last accessed 14 June 2017.

Arup-C40. 2016. *Deadline 2020: How cities will get the job done*. http://www.c40.org/researches/deadline-2020. Last accessed 14 June 2017.

Arup-Rockefeller. 2016. *City Resilience Index*. https://www.arup.com/perspectives/themes/cities/city-resilience-index. Last accessed 5 July 2017.

Atkins PLC. 2012. *Future proofing cities: Risks and opportunities for inclusive urban growth in developing countries*. Report published in collaboration with

UKAid/DFID and the Development Planning Unit (DPU), University College London. Last accessed 26 Feb 2013 at http://futureproofingcities.com/executive_summary.html.

Ayers, J.M., S. Huq, A.M. Faisal, and S.T. Hussain. 2014. Mainstreaming climate change adaptation into development: A case study of Bangladesh. *Wiley Interdisciplinary Reviews: Climate Change* 5 (1): 37–51.

Bäckstrand, K., and J.W. Kuyper. 2017. The democratic legitimacy of orchestration: The UNFCCC, non-state actors, and transnational climate governance. *Environmental Politics* 26 (4): 764–788.

Bäckstrand, Karin, Jonathan W. Kuyper, Björn-Ola Linnér, and Eva Lövbrand. 2017. Non-state actors in global climate governance: From Copenhagen to Paris and beyond. *Environmental Politics* 26 (4): 561–579.

Beck, U. 1992. *Risk society: Towards a new modernity*. London: Sage.

Betsill, Michele, and Harriet Bulkeley. 2004. Transnational networks and global environmental governance: The cities for climate protection program. *International Studies Quarterly* 48 (2): 471–493.

Betsill, M., N.K. Dubash, M. Paterson, et al. 2015. Building productive links between the UNFCCC and the broader global climate governance landscape. *Global Environmental Politics* 15 (2): 1–10.

Birkmann, J., M. Garschagen, F. Kraas, and N. Quang 2010. Adaptive urban governance: New challenges for the second generation of urban adaptation strategies to climate change. *Sustainability Science* 5 (2): 185–206.

Bouteligier, Sofie. 2015. Multinational companies and urban climate governance: Market making or successful policy innovation? In *The urban climate challenge: Rethinking the role of cities in the global climate regime*, ed. Craig Johnson, Noah Toly, and Heike Schroeder, 82–100. New York and London: Routledge Press.

Boyd, Emily, Aditya Ghosh, and Max Boykoff. 2015. Climate change adaptation in Mumbai, India. In *The urban climate challenge: Rethinking the role of cities in the global climate regime*, ed. Craig Johnson, Noah Toly, and Heike Schroeder, 139–155. New York and London: Routledge Press.

Brenner, Neil. 2013. Theses on urbanization. *Public Culture* 25 (1): 85–114. doi:10.1215/08992363-1890477.

Brenner, Neil, and Christian Scmid. 2013. The 'urban age' in question. *International Journal of Urban and Regional Research* 38 (3): 731–755. http://dx.doi.org/10.1111/1468-2427.12115.

Brenner, Neil, and N. Theodore. 2002. Cities and the geographies of 'actually-existing neo-liberalism'. *Antipode* 34 (3): 349–379.

Brooks, N., N. Grist, and K. Brown. 2009. Development Futures in the Context of Climate Change: Challenging the Present and Learning from the Past. *Development Policy Review* 27 (6): 741–765.

Bulkeley, H. 2010. Cities and the governing of climate change. *Annual Review of Environment and Resources* 35: 229–253.

Bulkeley, H., and V. Castan Broto. 2012. A survey of urban climate change experiments in 100 cities. *Global Environmental Change* 23 (1): 92–102.

Bulkeley, Harriet, Vanesa Castan Broto, and A.S. Edwards Gareth. 2015. *An urban politics of climate change*. London and New York: Routledge.

Burton, I. 2004 [2009]. Climate change and the adaptation deficit. In *The earthscan reader on adaptation to climate change*, ed. E.L.F. Schipper and I. Burton, 89–98. London: Earthscan.

Caravani, A., S. Barnard, S. Nakhooda, and L. Schalatek. 2014. Climate finance thematic briefing: Adaptation finance. *Climate Finance Fundamentals #3*, London: Overseas Development Institute. http://www.odi.org/sites/odi.org.uk/files/odi-assets/publications-opinion-files/9325.pdf. Last accessed 15 Feb 2015.

Carmin, J., I. Anguelovski, and D. Roberts. 2012. Urban climate adaptation in the global south: Planning in an emerging policy domain. *Journal of Planning Education and Research* 32 (1): 18–32.

Castan Broto, V. 2017. Urban governance and the politics of climate change. *World Development*. http://dx.doi.org/10.1016/j.worlddev.2016.12.031.

Christensen, C.M., D. Wang, and D. van Bever. 2013. Consulting on the cusp of disruption. *Harvard Business Review*, 2013. https://hbr.org/2013/10/consulting-on-the-cusp-of-disruption. Last accessed 23 June 2017.

Chu, E., I. Anguelovski, and J. Carmin. 2015. Inclusive approaches to urban climate adaptation planning and implementation in the Global South. *Climate Policy*, 16 (3): 1–21.

Chu, E., I. Anguelovski, and D. Roberts. 2017. Climate adaptation as strategic urbanism: Assessing opportunities and uncertainties for equity and inclusive development in cities. *Cities* 60: 378–387.

Davidson, K., and B. Gleeson. 2015. Interrogating urban climate leadership: Toward a political ecology of the C40 network. *Global Environmental Politics* 15 (4): 21–38.

Falkner, R. 2016. The Paris Agreement and the new logic of international climate politics. *International Affairs* 92 (5): 1107–1125.

Florida, R. 2017. *The new urban crisis: How our cities are increasing inequality, deepening segregation, and failing the middle class—And what we can do about it.* New York: Basic Books.

Folke, C. 2006. Resilience: The emergence of a perspective for social-ecological systems analysis. *Global Environmental Change* 16: 253–267.

Folke, C., T. Hahn, P. Olsson, and J. Norberg. 2005. Adaptive governance of social-ecological systems. *Annual Review of Environment and Resources* 30: 441–473.

Forsyth, T. 2013. Community-based adaptation: A review of past and future challenges. *WIREs Clim Change* 4: 439–446. doi:10.1002/wcc.231.

Gordon, David J. 2013. Between local innovation and global impact: Cities, networks, and the governance of climate change. *Canadian Foreign Policy Journal* 19 (3): 288–307. doi:10.1080/11926422.2013.844186.

Gordon, D.J. 2016. The politics of accountability in networked urban climate governance. *Global Environmental Politics* 16 (2): 82–100.

Gordon, D., and C. Johnson. 2017. The orchestration of global urban climate governance: Conducting power in the post-Paris climate regime. *Environmental Politics* 26 (4): 694–714.

Gordon, David, and Michele Acuto. 2015. If cities are the solution, what are the problems? The promise and perils of urban climate leadership. In *The urban climate challenge: Rethinking the role of cities in the global climate regime*, ed. Craig Johnson, Noah Toly, and Heike Schroeder, 63–81. New York and London: Routledge Press.

Gore, Christopher. 2015. Climate change adaptation and African cities: Understanding the impact of government and governance on future action. In *The urban climate challenge: Rethinking the role of cities in the global climate regime*, ed. Craig Johnson, Noah Toly, and Heike Schroeder, 205–226. New York and London: Routledge Press.

Grundmann, Reiner. 2013. 'Climategate' and the scientific ethos. *Science Technology Human Values* 38: 67. doi:10.1177/0162243911432318.

Haque, A.N., D. Dodman, and M.M. Hossain. 2014. Individual, communal and institutional responses to climate change by low-income households in Khulna, Bangladesh. *Environment and Urbanization* 26 (1): 112–129.

Hallegatte, S., C. Green, R.J. Nicholls, and J. Corfee-Morlot. 2013. Future flood losses in major coastal cities. *Nature climate change* 3(9): 802–806.

Hanson, S., R. Nicholls, N. Ranger, S. Hallegatte, J. Corfee-Morlot, C. Herweijer, and J. Chateau. 2011. A global ranking of port cities with high exposure to climate extremes. *Climatic change* 104(1): 89–111.

Harvey, D. 2006. Neo-liberalism as creative destruction. *Geografiska Annaler: Series B, Human Geography* 88 (2): 145–158.

Heltberg, R., P.B. Siegel, and S.L. Jorgensen. 2009. Addressing human vulnerability to climate change: Toward a 'no-regrets' approach. *Global Environmental Change* 19 (1): 89–99.
Hickmann, T. 2016. *Rethinking authority in global climate governance: How transnational climate initiatives relate to the international climate regime*. London and New York: Routledge Press.
Hochstetler, K., and M. Milkoreit. 2013. Emerging powers in the climate negotiations: shifting identity conceptions. *Political Research Quarterly* 67 (1): 224–235.
Hodson, M., F.W. Geels, and A. McMeekin. 2017. Reconfiguring urban sustainability transitions, analysing multiplicity. *Sustainability* 9 (2): 299.
Hoffmann, Matthew J. 2011. *Climate governance at the crossroads experimenting with a global response after Kyoto*. Oxford: Oxford University Press.
Holling, C.S. 2001. Understanding the complexity of economic, ecological, and social systems. *Ecosystems* 5 (4): 390–405.
Hoppe, R., A. Wesselink, and R. Cairns. 2013. Lost in the problem: The role of boundary organisations in the governance of climate change. *WIREs Clim Change* 4: 283–300. doi:10.1002/wcc.225.
Hurrell, A., and S. Sengupta. 2012. Emerging powers, North–South relations and global climate politics. *International Affairs* 88 (3): 463–484.
ICLEI. 2015. Carbonn climate registry: 5 year overview report: 2010–2015. http://www.carbonn.org. Last downloaded 10 Sep 2016.
Johnson, Craig. 2017. Holding polluting countries to account for climate change: Is 'loss and damage' up to the task? *Review of Policy Research* 34 (1): 50–67.
Johnson, Craig, Noah Toly, and Heike Schroeder. 2015. Urban resilience, low carbon governance and the global climate regime. In *The urban climate challenge: Rethinking the role of cities in the global climate regime*, 3–23. New York and London: Routledge Press.
Johnson, Craig, Iftekharul Haque, Yvonne Su, and Kristy May. 2017. Building urban climate resilience in Vietnam and Bangladesh. In *Building a climate resilient economy and society: Challenges and opportunities*, ed. K.N. Ninan and M. Inoue, 183–195. London: Edward Elgar.
Johnson, C., I. Haque, Y. Su, and K. May. (unpublished). *Building a climate for urban resilience: A cross-national comparison of urban climate adaptation in 17 Asian countries*. Department of Political Science, University of Guelph.
Kennedy, C. 2011. *The evolution of great world cities: Urban wealth and economic growth*. Toronto: University of Toronto Press.
Keohane, Robert, and David Victor. 2011. The regime complex for climate change. *Perspective on Politics* 9: 7–23.

Keohane, Robert and David Victor. 2016. Cooperation and discord in global climate policy. *Nature Climate Change*. 9 May 2016. doi:10.1038/NCLIMATE2937.

Lee, Taedong. 2015. *Global cities and climate change: The translocal relations of environmental governance*. New York: Routledge.

Leichenko, Robin. 2011. Climate change and urban resilience. *Current Opinion in Environmental Sustainability* 3: 164–168.

McCann, E. 2011. Urban policy mobilities and global circuits of knowledge: Toward a research agenda. *Annals of the Association of American Geographers* 101 (1): 107–130.

McDonald, Robert I., Pamela Green, Deborah Balk, Balazs M. Fekete, Carmen Revenga, Megan Todd, et al. 2011. Urban growth, climate change, and freshwater availability. *Proceedings of the National Academy of Sciences* 108 (15): 6312–6317.

McGranahan, G., D. Balk, and B. Anderson. 2007. The rising tide: Assessing the risks of climate change and human settlements in low elevation coastal zones. *Environment and Urbanization* 19 (1): 17–37.

Mehrotra, S., C. Rosenzweig, W.D. Solecki, C.E. Natenzon, A. Omojola, R. Folorunsho, J. Gilbride. 2011. Cities, disasters and climate risk. In *Climate change and cities: First assessment report of the urban climate change research network*, ed. C. Rosenzweig, W.D. Solecki, S.A. Hammer, and S. Mehrotra, 15–42. Cambridge: Cambridge University Press.

Moser, S., and A.L. Luers. 2008. Managing climate risks in California: The need to engage resource managers for successful adaptation to change. *Climatic Change* 87 (Suppl. 1): S309–S322.

Munslow, B., and T. O'Dempsey. 2011. Globalisation and climate change in Asia: The urban health impact. *Third World Quarterly* 8 (31): 1339–1356.

Nordhaus, W.D. 2007. A review of the stern review on the economics of climate. *Journal of Economic Literature* 45 (3): 686–702. doi:10.1257/jel.45.3.686.

Olsson, P., L.H. Gunderson, S.R. Carpenter, P. Ryan, L. Lebel, C. Folke, and C.S. Holling. 2006. Shooting the rapids: Navigating transitions to adaptive governance of social-ecological systems. *Ecology and Society* 11: 18.

Pahl-Wostl, C. 2009. A conceptual framework for analysing adaptive capacity and multi-level learning processes in resource governance regimes. *Global Environmental Change* 19: 354–365.

Perrow, C. 1984. *Normal accidents: Living with high risk technologies*. Princeton: Princeton University Press.

Perrow, C. 2011. *The next catastrophe: Reducing our vulnerabilities to natural, industrial, and terrorist disasters*. Princeton: Princeton University Press.

Revi, A., D. Satterthwaite, F. Aragón-Durand, J. Corfee-Morlot, R.B.R. Kiunsi, M. Pelling, D.C. Roberts, and W. Solecki. 2014. Urban areas. In *Climate Change 2014: Impacts, Adaptation, and Vulnerability. Part A: Global and Sectoral Aspects. Contribution of Working Group II to the Fifth Assessment Report of the Intergovernmental Panel on Climate Change*, ed. Field, C.B., V.R. Barros, D.J. Dokken, K.J. Mach, M.D. Mastrandrea, T.E. Bilir, M. Chatterjee, K.L. Ebi, Y.O. Estrada, R.C. Genova, B. Girma, E.S. Kissel, A.N. Levy, S. MacCracken, P.R. Mastrandrea, and L.L. White, 535–612. Cambridge, United Kingdom and New York, NY, USA: Cambridge University Press.

Ribot, J. 2014. Cause and response: vulnerability and climate in the Anthropocene. *Journal of Peasant Studies* 41 (5): 667–705.

Riddell, R. 2008. *Does Foreign Aid Really Work*? Oxford: Oxford University Press.

Roberts, E., Kees van der Geest, K. Warner, and S. Andrei. 2014. *Loss and damage: When adaptation isn't enough*. UNEP Global Environmental Alert Service. http://www.lossanddamage.net/download/7292.pdf. Last downloaded 13 February 2015.

Roberts, K. 2016. Climate change will be among KPIs for smart cities by 2020, says Gartner. *Internet of Business*. 11 Nov 2016. https://internetofbusiness.com/climate-change-smart-cities-gartner/. Last accessed 23 June 2017.

Romero-Lankao, P., and D. Dodman. 2011. Cities in transition: Transforming urban centers from hotbeds of GHG emissions and vulnerability to seed-beds of sustainability and resilience. *Current Opinion in Environmental Sustainability* 3: 113–120.

Romero-Lankao, P., and Hua Qin. 2011. Conceptualizing urban vulnerability to global climate and environmental change. *Current Opinion in Environmental Sustainability* 3: 142–149.

Romero-Lankao, P., D.M. Gnatz, O. Wilhelmi, and M. Hayden. 2016. Urban sustainability and resilience: From theory to practice. *Sustainability* 8 (12): 1224.

Rosenzweig, C., W.D. Solecki, S.A. Hammer, and S. Mehrotra (Eds.). 2011. *Climate change and cities: First assessment report of the urban climate change research network*. Cambridge University Press.

Rosenzweig, C., and W. Solecki. 2014. Hurricane Sandy and adaptation pathways in New York: Lessons from a first-responder city. *Global Environmental Change* 28: 395–408. http://dx.doi.org/10.1016/j.gloenvcha.2014.05.003.

Rosenzweig C., W. Solecki, P. Romero-Lankao, S. Mehrotra, S. Dhakal, T. Bowman, and S. Ali Ibrahim. 2015. *ARC3.2 Summary for city leaders urban climate change research network*. New York: Columbia University.

Roy, M., D. Hulme, and F. Jahan. 2013. Contrasting adaptation responses by squatters and low-income tenants in Khulna, Bangladesh. *Environment and Urbanization* 25 (1): 157–176.

Sassen, S. 2015. Bringing cities into the global climate framework. In *The urban climate challenge: Rethinking the role of cities in the global climate regime*, ed. Craig Johnson, Noah Toly, and Heike Schroeder, 24–36. New York and London: Routledge Press.

Sassen, S., and N. Dotan. 2011. Delegating, not returning, to the biosphere: How to use the multi-scalar and ecological properties of cities. *Global Environmental Change* 21: 823–834.

Satterthwaite, D., S. Huq, H. Reid, M. Pelling, P. Romero Lankao. 2007. Adapting to climate change in urban areas: The possibilities and constraints in low- and middle-income nations. *IIED Human Settlements Discussion Paper*. Last accessed 23 Mar 2012 at http://pubs.iied.org/pdfs/10549IIED.pdf.

Satterthwaite, D., G. McGranahan, and C. Tacoli. 2010. Urbanization and its implications for food and farming. *Philosophical Transactions of the Royal Society of London B: Biological Sciences* 365: 2809–2820.

Satterthwaite, D., and S. Bartlett. 2017. Editorial: The full spectrum of risk in urban centres: changing perceptions, changing priorities. *Environment and Urbanization* 29 (1): 3–14. doi:10.1177/0956247817691921.

Schalatek, L., and N. Bird. 2014. The Principles and Criteria of Public Climate Finance - A Normative Framework. *Climate Finance Fundamentals #1*, London: Overseas Development Institute. http://www.odi.org/sites/odi.org.uk/files/odi-assets/publications-opinion-files/9296.pdf. Last accessed 15 Feb 2015.

Sennett, R. 2012. No one likes a city that's too smart. *The Guardian*. 4 Dec 2012. https://www.theguardian.com/commentisfree/2012/dec/04/smart-city-rio-songdo-masdar. Last accessed 23 June 2017.

Serdeczny, O., E. Waters, and S. Chan. 2016. *Non-economic loss and damage in the context of climate change*. German Development Institute Discussion Paper 3/2016. Bonn: Deutsches Institut für Entwicklungspolitik.

Shi, Linda. 2016. Adapting to climate change in cities may require a major rethink. *New Security Beat,* a blog of the Wilson Center's Environmental Change and Security Program. https://www.newsecuritybeat.org/2016/02/adapting-climate-change-cities-require-major-rethink/?mkt_tok=eyJpIjoi

WmpGaU4yRXdabVUwTTJNMiIsInQiOiJiVEVxNldRZmxkaVJxRGQrbFlQbXVOQk9CRjU3VitEU0Uyczlkajdsd0Z4NTF6dEdhXC9nUUx6S0p3cWJ6b2hRcVgwSFBlNGV2K2ZDaXRRMUFBWFRJXC9aKzJzb0NLMGhQSkVtXC9TTlc0TlwvRlZWYUZzWkJuV3RMUzhkV2NveVFpTEYifQ%3D%3D. Last accessed 10 May 2017.

Shi, L., E. Chu, I. Anguelovski, A. Aylett, J. Debats, K. Goh, ... and J.T. Roberts. 2016. Roadmap towards justice in urban climate adaptation research. *Nature Climate Change* 6 (2): 131–137.

Shue, H. 2010. Subsistence Emissions and Luxury Emissions. In *Climate ethics: Essential readings*, ed. Gardiner et al., 200–214. Oxford: Oxford University Press.

Smit B., and J. Wandel. 2006. Adaptation, adaptive capacity and vulnerability. *Global Environmental Change* 16 (3): 282–292.

Sovacool, B.K. 2011. Hard and soft paths for climate change adaptation. *Climate Policy* 11 (4): 1177–1183. doi:10.1080/14693062.2011.579315.

Sovacool, B.K., B.O. Linnér, and M.E. Goodsite. 2015. The political economy of climate adaptation. *Nature Climate Change* 5 (7): 616–618.

Stevenson, Hayley, and John Dryzek. 2014. *Democratizing global climate governance*. Cambridge: CUP.

UN/DESA (United Nations/Department of Economic and Social Affairs). 2012. *World urbanization prospects: The 2011 revision*. Downloaded 23 Jan 2013 at http://esa.un.org/unpd/wup/pdf/WUP2011_Highlights.pdf.

UNEP. 2016. *The adaptation finance gap report 2016*. Nairobi, Kenya: United Nations Environment Programme (UNEP). Retrieved 24 May 2017 from http://www.unep.org/adaptationgapreport/sites/unep.org.adaptationgapreport/files/documents/agr2016.pdf.

UNFCCC. 2015a. http://unfccc.int/adaptation/workstreams/loss_and_damage/items/8134.php. Last accessed 9 Feb 2015.

UNFCCC. 2015b. Adoption of the Paris Agreement Conference of the Parties 21st Session, Paris, 30 November–11 December 2015. http://unfccc.int/resource/docs/2015/cop21/eng/l09r01.pdf. Last accessed 18 Dec 2015.

UNISDR (UN International Office for Disaster Risk Reduction). 2012. *Making cities resilient report 2012*. Retrieved 26 Feb 2013 from www.unisdr.org.

van der Ven, H., S. Bernstein, and M. Hoffmann. 2017. Valuing the contributions of nonstate and subnational actors to climate governance. *Global Environmental Politics* 17 (1): 1–20.

Wachsmuth, D., D.A. Cohen, and H. Angelo. 2016. Expand the frontiers of urban sustainability. *Nature* 536 (7617): 391–393.

Warner, K., K. van der Geest, S. Kreft, S. Huq, S. Harmeling, K. Kusters, and A. De Sherbinin. 2012. *Evidence from the frontlines of climate change: Loss and damage to communities despite coping and adaptation Loss and Damage in Vulnerable Countries Initiative*. Policy Report. Report No. 9. United Nations University Institute for Environment and Human Security (UNU-EHS), Bonn. Accessed 12 April 2014.

Warner, K., K. van der Geest, and S. Kreft, 2013. *Pushed to the limits: Evidence of climate change-related loss and damage when people face constraints and limits to adaptation*. Report No. 11. United Nations University Institute of Environment and Human Security (UNU-EHS), Bonn. Accessed 12 April 2014.

Widerberg, O., and J. Stripple. 2016. The expanding field of cooperative initiatives for decarbonization: A review of five databases. *Wiley Interdisciplinary Reviews: Climate Change* 7: 486–500.

Wisner, B., P. Blaikie, T. Cannon, and I. Davis. 2004. *At risk: Natural hazards, people's vulnerability and disasters*. London and New York: Routledge.

World Bank. 2010a. *Cities and climate change: An urgent agenda*. Washington, DC: World Bank. http://siteresources.worldbank.org/INTUWM/Resources/340232-1205330656272/CitiesandClimateChange.pdf. Retrieved 24 May 2017.

World Bank. 2010b. *The economics of adaptation to climate change*. Washington, DC: World Bank. https://siteresources.worldbank.org/EXTCC/Resources/EACC_FinalSynthesisReport0803_2010.pdf. Retrieved 24 May 2017.

Zommers, Z. D. Wrathall, and K. van der Geest. 2014. *Loss and Damage to Ecosystem Services*. UNU-EHS Working Paper Series, No. 2. Bonn: United Nations University Institute of Environment and Human Security (UNU-EHS).

5

Taking It to the Streets (and Beyond): The Power of Cities in Global Climate Politics

Abstract This final chapter concludes the analysis by highlighting the critical ways in which city engagement in global climate politics affects our understanding of international power. The implications for conceptualizing and theorizing the power of cities are considered, as are the methodologies that may be used in exploring and explaining the changing power of cities in global environmental governance regimes.

Keywords Urban power · Transnational climate politics
Low carbon futures · New directions in theory · Research and analysis

5.1 Introduction

The last two chapters have considered the ways in which cities and city-networks have affected and projected power in global climate politics. One is the idea that they are stepping into a void created by intransigent national governments. A second is that cities and city-networks are articulating, negotiating and demonstrating a need for recognition and assistance that may facilitate adaptation at the urban scale.

C.A. Johnson, *The Power of Cities in Global Climate Politics*, Cities and the Global Politics of the Environment, https://doi.org/10.1057/978-1-137-59469-3_5

This final chapter now concludes the analysis by highlighting some of the critical ways in which city engagement in global climate politics affects our understanding of international power. The implications for conceptualizing and theorizing the power of cities are considered, as are the methodologies that may be used in exploring and explaining the changing power of cities in global environmental governance regimes.

5.2 Climate Governance in the Urban Age: Coming of Age or Turning the Page?

A key question that emerges from the preceding analysis is whether and to what extent the involvement of cities and city-networks in global climate politics constitutes a new form of power that challenges—or even renders obsolete—our traditional understanding of international power (Curtis 2014, 2016; Davis 2016). On the one hand, cities are clearly occupying spaces (e.g. attending international conferences, articulating foreign policy, setting international policy agendas) that have until recently been the exclusive domain of nation-states (Tavares 2016; Barber 2013, 2017). On the other, their ability to engage in multilateral processes (e.g. NAZCA) remains constrained by a Westphalian model of international order.

Understanding the changing nature of city power necessarily entails making sense of the historical and ideological forces that have bearing on the power and autonomy of the nation-state (Herrschel and Newman 2017). As noted in Chap. 2, the rise of cities in global climate politics can be usefully associated with the "crisis of statism" that manifested itself primarily during the neoliberal reform era of the 1980s (Brenner 2013; Curtis 2014, 2016; Davidson and Gleeson 2015). In spite of the (apparent) neo-populist backlash that has recently occurred in Western Europe and the USA, neoliberal models continue to influence the ways in which cities and states approach the problem of urban climate governance. In the case of climate mitigation, the crisis of statism has manifested itself most significantly in the form of reneging on international agreements, reducing or eliminating funding for national climate research, stifling the production of alternative energy and undermining the autonomy of cities to reduce emissions within their jurisdictions (Davidson and Gleeson 2015; Romero-Lankao et al. 2015; Aykut 2016). With adaptation, state retrenchment has entailed privatizing and eliminating essential urban services, facilitating private investment (particularly in the property sector) and rejecting traditional state-led models of urban planning and development (Davidson and Gleeson 2015; Romero-Lankao et al. 2015; Romero-Lankao et al. 2016; Shi et al. 2016).

To say that all cities have engaged in this kind of neoliberalism is of course an overstatement, but a critical point is that transnational city governance has occurred in a context of state retrenchment that has allowed—and in many ways necessitated—a more ambitious role on the part of the sub- and non-state actors.

A second and related issue concerns the power and autonomy of transnational city-networks. As noted in Chaps. 3 and 4, city-networks have clearly established new forms of coordination, orchestration and governance that are now playing an important role in the contemporary climate regime. However, participation and engagement in city-networks have been uneven, reflecting a preference on the part of networks like *100 Resilient Cities* and the C40 for large and well-connected cities, as well as a lack of funds and resources (cf. Hickmann 2016). Moreover, the impacts of city-network governance on local policy remain unclear, highlighting critical questions for future policy and research (see below).

A third theme concerns the power of corporations—and corporate models of climate governance—in shaping the agendas and ideals that cities, city-networks and nation-states are now pursuing in the name of climate adaptation and mitigation. From the preceding, we can infer that MNCs like Siemens and Arup have come to play a critical role in developing and disseminating the strategies that city-networks are now using to achieve their targets for mitigating and adapting to climate change (Bouteligier 2015; Gordon and Acuto 2015; Acuto and Rayner 2016; Gordon and Johnson 2017). Although this influence is most readily apparent in the case of the C40, corporate models can also be found in the inventories and vulnerability assessments of ICLEI, the Global Compact of Mayors and the *100 Resilient Cities*.

A fourth question is whether the performative power of national governments, multilateral institutions, transnational city-networks and multinational corporations is having a discernible effect on the ways in which cities incorporate and adopt strategies for mitigation and adaptation at the urban scale. As noted in Chaps. 3 and 4, cities and city-networks have become increasingly involved in disseminating standards, metrics, methodologies and reporting rubrics for observing, evaluating and influencing the ways in which cities are mitigating and adapting to climate change. To the extent that these norms are actually being adopted and applied by cities, the emergent orchestration of network governance suggests a new form of soft power that has bearing on our understanding of global climate politics. However, questions remain about whether and to what extent cities are actually adopting these metrics and standards (see below).

A fifth question is whether the combination of city, corporate and network governance amounts to a consolidation or a fragmentation of power within global climate politics. As noted in Chap. 2, the UNFCCC is often portrayed as a weak regime that lacks an ability to enforce compliance on the part of member states (e.g. Keohane and Victor 2011, 2016). Although cities are often portrayed as an alternative to the multilateral system, it is worth noting here too that city-networks possess but a loose set of instruments and incentives (e.g. metrics, guidelines and financial inducements) that they can use in eliciting voluntary contributions from member cities. In this sense, the ability of city-networks to advance a coherent climate policy agenda is highly dependent upon the willingness and ability of cities to institutionalize their plans, policies and metrics at the urban scale (a point we explore in more detail below).

A sixth and final set of questions concerns the cumulative impact of city and city-network engagement on reducing emissions and vulnerability to climate change. Over the last decade, cities and city-networks have become increasingly involved in pursuing an ambitious climate policy agenda, including most recently a series of strategies for achieving the Paris target of 1.5 C (Arup-C40 2015a, b, 2016). At the same time, systematic evaluations of city actions and commitments are thin (Widerberg and Stripple 2016), suggesting ample room for future research (see below). Whether cities and city-networks will make a difference in terms of reducing emissions and vulnerability to climate change will depend on the ability of municipalities, businesses and local communities to devise new ways of governing and adapting to social and environmental stress. But it will also depend on the ability of cities to support policy initiatives that work with a wider range of state and non-state actors whose interests, actions and institutions have important bearing on the ability to engage in mitigation and adaptation at wider scales of interaction.

5.3 Counter Currents: On the Limits of City Power

Contradicting the apparent groundswell of enthusiasm that is now being associated with the involvement of cities, city-networks and other sub- and non-state actors in global climate politics are a number of questions about the future of the city and city-network power in global climate politics. First, to what extent are cities and city-networks able to pursue an agenda that is independent of multinational corporations, international institutions and nation-states? Second, how sustainable are these agendas? And third, what are the limits of city power?

As the preceding suggests, cities and city-networks have become increasingly engaged in establishing rules, norms and metrics for reducing emissions and vulnerability to climate change. However, cities are also embedded in political and economic systems that conceivably constrain their ability to act on the international stage. Particularly important in this regard is the extent to which the agency of cities (and city leaders) is contingent upon the power of the nation-state. Diane Davis (2016) poses this question in the following way:

> The nation-state emerged as the dominant form of sovereignty and governance in modern times because it was better equipped than city-states or empires to marshal financial resources, armed support and political legitimacy to prevail in war. Whether the nation-state will prove a historic aberration or whether violent suppression of would-be city-states is

> possible, now that technological prowess has replaced standing armies as the determining factor of military might, remains to be seen.

In a context of hyper-globalization that has involved some truly revolutionary changes in communication, economic organization and human mobility, cities have acquired new forms of wealth, influence and power by developing, innovating and adopting new modes of production, consumption and exchange (Harvey 2008; Brenner 2013; Curtis 2014, 2016). At the same time, their ability to thrive and survive in the global economy remains highly dependent upon the regulatory, financial and military power of nation-states (Harvey 2008; Brenner 2013). Beyond the administrative, regulatory and legislative powers that national governments have at their disposal is the cumulative impact of national policy trends and processes on mitigation and adaptation. Despite American intransigence on climate change, current projections suggest that India and China will both exceed their Paris commitments, largely as a result of national policies on eliminating coal-fired electricity.[1]

A second and critical factor is the relative wealth and prosperity of cities and city-networks. As noted in Chaps. 3 and 4, city-networks have tended to coalesce around large "global cities", whose wealth and influence have been instrumental in driving a global agenda on climate change. In this sense, the leadership of affluent cities and city leaders has been crucial, raising questions about the ability of city-networks to develop autonomous plans for mitigating and adapting to climate change. As noted in Chaps. 3 and 4, city-networks like ICLEI and the C40 have become highly dependent upon the contributions of private corporations and corporate foundations, such as Bloomberg Philanthropies, the Rockefeller Foundation and Arup, suggesting a possible limitation on the ability of city-networks to develop an autonomous climate policy agenda (see below).

A third and related factor concerns the extent to which cities and national governments are able to institutionalize mitigation and adaptation into local urban policy agendas. From the preceding, we can infer

[1]http://climateactiontracker.org/news/278/China-India-slow-global-emissions-growth-Trumps-polices-will-flatten-US-emissions.html. Last accessed 3 July 2017.

that many cities have adopted action plans and agendas that have been disseminated by cities, city-networks, MNCs and the UNFCCC. However, institutionalizing climate change into local policy also entails mainstreaming and incorporating mitigation and adaptation norms into operational policy and planning (Aylett 2015; Burch et al. 2015). Although many cities have adopted ambitious climate policy commitments, their ability to meet these objectives is often constrained by competing policy norms, such as the demand for affordable housing, sustainable livelihood, convenient shopping and reliable energy. Crucially, not all of these will be consistent with a resilient, low-carbon future.

A fourth and final point is that cities are also sites of protest and resistance, in which the ability to challenge and contest the terms of global resource distribution—including the use and abuse of the atmospheric commons—creates new sources of instability and democracy. As noted in Chap. 1, theories of sustainability transitions often entail disrupting the dominant sociotechnical and socio-ecological systems that underlie the creation of urban goods and services, the consumption of resources and the production of waste and pollution (Geels 2004; Geels and Raven 2006; Geels and Schot 2007; Bulkeley et al. 2015). Whether cities and city-networks have the capacity to disrupt or "destabilize" existing patterns of resource consumption and production, important comparative questions can be raised about the extent to which local and national authorities are able to challenge the prevailing economic and political forces that are driving greenhouse gas emissions while at the same time addressing the factors that perpetuate poverty, inequality and vulnerability at the urban scale.

5.4 The Road Ahead: New Directions in Urban Climate Research

The preceding analysis suggests a number of theoretical and empirical questions that have important bearing on the future of urban climate policy, research and scholarship. One concerns the extent to which and ways in which cities have adopted the standards, metrics and methodologies that are being disseminated by transnational city-networks, multinational corporations and the UNFCCC. As noted in Chap. 3, networks of international researchers are undertaking new efforts to standardize and evaluate city- and sub-national reporting on climate change. Two important recent examples of this kind include the Initiative for Climate Action Transparency (http://www.climateactiontransparency.org) and Galvanizing the Groundswell (http://www.climategroundswell.org/). Collaborative efforts such as these will undoubtedly provide a systematic body of information that can be used to understand whether and to what extent cities, city councils and local administrative bodies have adopted adaptation and mitigation norms into operational policy and practice. At the same time, more needs to be known about the extent to which the transnational norms and metrics that are being used to frame and evaluate city performance are consistent with local norms, interests and priorities, and whether city-networks and the UNFCCC are able to support local authorities and communities in developing the capacity for reporting and pursuing adaptation and mitigation agendas. Incorporating and assessing the needs of poor and politically marginal populations and cities remain a particularly important research priority (Chu et al. 2015; Shi et al. 2016).

Empirical questions about local adoption and inclusion raise wider theoretical questions about the nature of network orchestration, policy diffusion and political integration. As noted in Chap. 1, the power of cities and city-networks to orchestrate local actions and commitments can be usefully understood as a kind of soft power that operates alongside or *in the shadows of* more formal, multilateral arrangements, such as the Paris Climate Agreement. However, much remains to be known about the ways in which cities and city-networks are able to offer an alternative to the formal mechanisms of the UNFCCC. Particularly important in this regard is the extent to which cities and city-network are simply adopting or mimicking multilateral goals and policies, as opposed to pushing them into radically new ways of thinking about urban climate governance (e.g. Arup-C40 2016). Lacking an independent body of research that can be used to verify the extent to which cities have actually institutionalized and implemented city-network norms, plans and policies, it is difficult to ascertain the impact of network power. Similarly, more needs to be known about the ways in which and extent to which MNCs and corporate models of environmental governance are influencing the plans and policies that cities are now using to mitigate and adapt to climate change.

A third and related question concerns the politics of inclusion and exclusion. If we accept the notion that multilateral and transnational climate governance employs a kind of performative power that entails observing, comparing and orchestrating urban subjects, then important questions can be asked about the implications of being seen, compared and orchestrated. Underlying *CRAFT*, carbon*n* and the GPC is an assumption that conforming to a particular set of methodologies and standards will provide a means of connecting cities and urban populations with important financial, political and environmental regimes. However, as the preceding suggests, network participation and engagement have been uneven, raising empirical and theoretical questions about the conditions under which cities join and abide by network governance norms and procedures. As noted in Chap. 4, the insurance industry is now putting additional pressure on cities to alter their zoning restrictions and by-laws as a way of preventing housing and construction in areas that are prone to flooding and erosion. However, these are also areas that happen to be particularly important for poor and politically marginal groups and communities, raising important questions about the ways in which cities and local communities can balance the risks and vulnerabilities of a changing climate with the needs of the urban poor.

A related question concerns the changing nature of international climate policy and finance. As noted in Chaps. 3 and 4, multilateral funding for urban mitigation and adaptation remains well below what is realistically needed for reducing emissions and vulnerability to climate change. The Green Climate Fund represents a new and important source of funding, but here too much more needs to be known about the ways in which multilateral, bilateral and private donors can build capacity and resilience at the urban scale. Particularly important is a better understanding of the kinds of policy options and trade-offs that are going to challenge local, national and international actors in the years ahead. Assuming the international community is able to mobilize $100 bn per year by 2020, significant questions remain about how the money can best be spent.

Looking forward, a final theme that is likely to have enormous bearing on the future of global climate politics is the extent to which cities and transnational city-networks are able to translate the financial, demographic and economic power that has been ascribed to them into more substantial political power on the international stage. Within the multilateral system, cities and city-networks remain highly dependent upon the acquiescence and support of multilateral institutions and nation-states. Beyond the UNFCCC, cities and city-networks represent an important alternative to the state-centred model, but their potential for disrupting and re-shaping the politics of global climate governance remains uncertain, suggesting ample scope for further research.

References

Acuto, M., and S. Rayner. 2016. City networks: Breaking gridlocks or forging (new) lock-ins? *International Affairs* 92 (5): 1147–1166.

Arup-C40. 2015a. *Climate action in megacities 3.0*. http://www.cam3.c40.org/images/C40ClimateActionInMegacities3.pdf. Last accessed 14 June 2017.

Arup-C40. 2015b. *Potential for climate action*. http://publications.arup.com/publications/p/potential_for_climate_action. Last accessed 14 June 2017.

Arup-C40. 2016. *Deadline 2020: How cities will get the job done*. http://www.c40.org/researches/deadline-2020. Last accessed 14 June 2017.

Aykut, S.C. 2016. Taking a wider view on climate governance: Moving beyond the 'iceberg', the 'elephant', and the 'forest'. *Wiley Interdisciplinary Reviews: Climate Change* 7 (3): 318–328.

Aylett, Alex. 2015. Relational agency and the local governance of climate change: International trends and an American exemplar. In *The urban climate challenge: Rethinking the role of cities in the global climate regime*, ed. Craig Johnson, Noah Toly, and Heike Schroeder, 156–180. New York and London: Routledge Press.

Barber, Benjamin R. 2013. *If mayors ruled the world: Dysfunctional nations, rising cities*. New Haven, CT: Yale University Press.

Barber, B. 2017. *Cool cities: Urban sovereignty and the fix for global warming*. New Haven: Yale University Press.

Bouteligier, Sofie. 2015. Multinational companies and urban climate governance: Market making or successful policy innovation? In *The urban climate challenge: Rethinking the role of cities in the global climate regime*, ed. Craig Johnson, Noah Toly, and Heike Schroeder, 82–100. New York and London: Routledge Press.

Brenner, Neil. 2013. Theses on urbanization. *Public Culture* 25 (1): 85–114. doi:10.1215/08992363-1890477.

Bulkeley, Harriet, Vanesa Castan Broto, and Gareth A.S. Edwards. 2015. *An urban politics of climate change*. London and New York: Routledge.

Burch, Sarah, Alison Shaw, Freya Kristensen, John Robinson, and Ann Dale. 2015. Urban climate governance through a sustainability lens: Exploring the integration of adaptation and mitigation in four British Columbian cities. In *The urban climate challenge: Rethinking the role of cities in the global climate regime*, ed. Craig Johnson, Noah Toly, and Heike Schroeder, 119–138. New York and London: Routledge Press.

Chu, E., I. Anguelovski, and J. Carmin. 2015. Inclusive approaches to urban climate adaptation planning and implementation in the global south. *Climate Policy* 16 (3): 1–21. http://dx.doi.org/10.1080/14693062.2015.1019822.

Curtis, S. 2014. The meaning of global cities: Rethinking the relationship between cities, states and international order. In *The power of cities in international relations*, ed. S. Curtis, 16–31. New York and London: Routledge.

Curtis, S. 2016. Cities and global governance: State failure or a new global order? *Millennium* 44 (3): 455–477.

Davidson, K., and B. Gleeson. 2015. Interrogating urban climate leadership: Toward a political ecology of the C40 network. *Global Environmental Politics* 15 (4): 21–38.

Davis, D. 2016. Can mayors actually rule the world? *Citiscope*. http://citiscope.org/habitatIII/commentary/2016/12/can-mayors-actually-rule-world?utm_source=Citiscope&utm_campaign=a10e62b5f0-Mailchimp_H3_2016_12_05&utm_medium=email&utm_term=0_ce992dbfef-a10e62b5f0-118068273. Last accessed 7 December 2016.

Geels, F. 2004. From sectoral systems of innovation to socio-technical systems: Insights about dynamics and change from sociology and institutional theory. *Research Policy* 33: 897–920.

Geels, F. and Raven, R. 2006. Non-linearity and expectations in niche-development trajectories: Ups and downs in dutch biogas development (1973–2003). *Technology Analysis & Strategic Management* 18 (3–4): 375–392. doi:10.1080/09537320600777143.

Geels, F. and J. Schot. 2007. Typology of sociotechnical transition pathways. *Research Policy* 36: 399–417.

Gordon, D., and C. Johnson. 2017. The orchestration of global urban climate governance: Conducting power in the post-Paris climate regime. *Environmental Politics* 26 (4): 694–714.

Gordon, David, and Michele Acuto. 2015. If cities are the solution, what are the problems? The promise and perils of urban climate leadership. In

The urban climate challenge: Rethinking the role of cities in the global climate regime, ed. Craig Johnson, Noah Toly, and Heike Schroeder, 63–81. New York and London: Routledge Press.
Harvey, D. 2008. The right to the city. *New Left Review* 53: 23–40.
Herrschel, T., and P. Newman. 2017. *Cities as international actors: Urban and regional governance beyond the nation state*. London: Palgrave/Macmillan.
Hickmann, T. 2016. *Rethinking authority in global climate governance: How transnational climate initiatives relate to the international climate regime*. London and New York: Routledge Press.
Keohane, Robert, and David Victor. 2011. The regime complex for climate change. *Perspective on Politics* 9: 7–23.
Keohane, Robert, and David Victor. 2016. Cooperation and discord in global climate policy. *Nature Climate Change*. 9 May 2016. doi:10.1038/NCLIMATE2937.
Romero-Lankao, Patricia, et al. 2015. Multilevel governance and institutional capacity for climate change responses in Latin American cities. In *The urban climate challenge: Rethinking the role of cities in the global climate regime*, ed. Craig Johnson, Noah Toly, and Heike Schroeder, 181–204. New York and London: Routledge Press.
Romero-Lankao, P., D.M. Gnatz, O. Wilhelmi, and M. Hayden. 2016. Urban sustainability and resilience: From theory to practice. *Sustainability* 8 (12): 1224.
Shi, L., E. Chu, I. Anguelovski, A. Aylett, J. Debats, K. Goh …, and J.T. Roberts. 2016. Roadmap towards justice in urban climate adaptation research. *Nature Climate Change* 6 (2): 131–137.
Tavares, R. 2016. *Paradiplomacy: cities and states as global players*. Oxford: Oxford University Press.
Widerberg, O., and J. Stripple. 2016. The expanding field of cooperative initiatives for decarbonization: A review of five databases. *Wiley Interdisciplinary Reviews: Climate Change* 7: 486–500.

References

ADB (Asian Development Bank). 2011. *Adapting to climate change: Strengthening the climate resilience of the water sector infrastructure in Khulna, Bangladesh*. Manila: ADB. www.adb.org. Last accessed 22 Aug 2015.

Aylett, Alex. 2013. The socio-institutional dynamics of urban climate governance: A comparative analysis of innovation and change in Durban (KZN, South Africa) and Portland (OR, USA). *Urban Studies* 50 (7): 1386–1402.

Bäckstrand, K., and E. Lövbrand. 2016. The road to Paris: Contending climate governance discourses in the post-Copenhagen era. *Journal of Environmental Policy & Planning*. doi:10.1080/1523908X.2016.1150777.

Bennett, C. 1991. What is policy convergence and what causes it? *British Journal of Political Science* 21 (2): 215–233.

Betsill, Michele, and Harriet Bulkeley. 2006. Cities and the multilevel governance of global climate change. *Global Governance* 12 (2): 141–159.

Betsill, Michele, and Harriet Bulkeley. 2007. Looking back and thinking ahead: A decade of cities and climate change research. *Local Environment* 12 (5): 447–456.

Chan, S., R. Falkner, M. Goldberg, and H. van Asselt. 2016. Effective and geographically balanced? An output-based assessment of non-state climate actions. *Climate Policy*. doi:10.1080/14693062.2016.1248343.

Collier, David. 2011. Understanding process tracing. *PS: Political Science and Politics* 44 (4): 823–830.

C.A. Johnson, *The Power of Cities in Global Climate Politics*, Cities and the Global Politics of the Environment, https://doi.org/10.1057/978-1-137-59469-3

COP21. 2015. *Paris agreement united nations framework convention on climate change, Paris, France*. http://unfccc.int/paris_agreement/items/9485.php. Last accessed 19 June 2016.

Dodman, David, Jane Bicknell, and David Satterthwaite. 2009. *Adapting cities to climate change: Understanding and addressing the development challenges*. London: Earthscan.

George, Alexander, and Andrew Bennett. 2004. *Case studies and theory development in the social sciences*. Cambridge, MA: MIT Press.

Hilton, I., and O. Kerr. 2017. The Paris agreement: China's 'new normal' role in international climate negotiations. *Climate Policy* 17 (1): 48–58. doi:10.1080/14693062.2016.1228521.

Hodson, Mike, and Simon Marvin. 2010. Urbanism in the anthropocene: Ecological urbanism or premium ecological enclaves? *City* 14 (3): 299–312.

IPCC (Intergovernmental Panel on Climate Change, WG2). 2014. *Chapter 8: Urban areas*. http://ipcc-wg2.gov/AR5/images/uploads/WGIIAR5-Chap8_FGDall.pdf. Last downloaded 5 Aug 2014.

Khare, Anshuman, and Terry Beckman (eds.). 2013. *Mitigating climate change: The emerging face of modern cities*. London: Springer.

McDonald, R.I., et al. 2011. Urban growth, climate change, and freshwater availability. *PNAS*. Published online before print March 28, 2011. doi: 10.1073/pnas.1011615108.

Moser, C., and D. Satterthwaite. 2008. *Towards pro-poor adaptation to climate change in urban centres of low- and middle-income countries*. London: Human Settlements Discussion Paper Series IIED.

Newman, Peter, Timothy Beatley, and Heather Boye. 2009. *Resilient cities: Responding to peak oil and climate change*. Washington: Island Press.

Poon, L. 2016. Mayors set a tight deadline to initiate climate action. *The Atlantic Citylab*. http://www.citylab.com/politics/2016/12/immediate-climate-action-falls-on-mayors/509168/?utm_source=nl__link3_120216. Last accessed 7 Dec 2016.

Sabatier, P.A., and C.M. Weilbe. 2007. The advocacy coalition framework: Innovations and clarifications. In *Theories of the policy process Boulder*, ed. P. Sabatier, 189–222. Boulder: Westview Press.

Sanyal, Bishwapriya, Lawrence J. Vale, and Cristina D. Rosan (eds.). 2012. *Planning ideas the matter: Livability, territoriality, governance, and reflective practice*. Cambridge: MIT Press.

Seto, K. 2011. Exploring the dynamics of migration to mega-delta cities in Asia and Africa: Contemporary drivers and future scenarios. *Global Environmental Change* 21 (Suppl. 1): S94–S107.

Seto, K., M. Fragkias, B. Guneralp, and M.K. Reilly. 2011. A meta-analysis of global urban land expansion. *PLoS One* 6 (8): 1–9.

Shah, Kalim U., Hari Bansha Dulal, Craig Johnson, and April Baptiste. 2013. Understanding livelihood vulnerability to climate change: Applying the livelihood vulnerability index in Trinidad and Tobago. *Geoforum* 47: 125–137.

Smith, A., A. Stirling, and F. Berkhout. 2005. The governance of sustainable socio-technical transitions. *Research Policy* 34: 1491–1510.

Solecki, W., K.C. Seto, D. Balk, A. Bigio, C.G. Boone, F. Creutzig, M. Fragkias, S. Lwasa, P. Marcotullio, P. Romero-Lankao, and T. Zwickel. 2015. A conceptual framework for an urban areas typology to integrate climate change mitigation and adaptation. *Urban Climate* 14: 116–137.

Swyngedouw, E. 1997. Power, nature, and the city: The conquest of water and the political ecology of urbanization in Guayaquil, Ecuador: 1880–1990. *Environment and Planning A* 29: 311–332.

Taylor, Peter. 2001. Specification of the world city network. *Geographical Analysis* 33 (2): 181–194.

Winfield, Mark. 2015. Implementing environmental policy in Canada. In *Canadian environmental policy and politics*, ed. D. VanNijnatten, 74-96. Toronto: Oxford University Press.

World Bank. 2015. *Urban population figures*. http://data.worldbank.org/indicator/SP.URB.TOTL.IN.ZS. Last accessed 28 Sep 2015.

Index

C.A. Johnson, *The Power of Cities in Global Climate Politics*, Cities and the Global Politics of the Environment, https://doi.org/10.1057/978-1-137-59469-3

FSC
www.fsc.org
MIX
Papier | Fördert
gute Waldnutzung
FSC® C083411

Kolibri 360 GmbH
Ferdinand-Jühlke-Straße 7
99095 Erfurt, Deutschland
produktsicherheit@kolibri360.de